INTERMEDIATE ALGEBRA AND THE GRAPHING CALCULATOR

A LEARNING RESOURCE

JAMES ANGELOS

Central Michigan University

ANITA M. JOHNSTON

DONAL B. STAAKE, JR.

Jackson Community College

SAUNDERS COLLEGE PUBLISHING
HARCOURT BRACE COLLEGE PUBLISHERS

Fort Worth Philadelphia San Diego New York Orlando San Antonio Austin
Montreal London Sydney Tokyo

Printed in the United States of America.

Angelos/Johnston/Staake, Jr.: Intermediate Algebra and the Graphing Calculator: A Learning Resource, to accompany INTERMEDIATE ALGEBRA: A TEXT/WORKBOOK, 4/E, by Charles P. McKeague

ISBN 0-03-010772-5

6 7 8 9 0 1 2 023 9 8 7 6 5 4

Preface

Graphing Calculator Investigations has been written to accompany *Intermediate Algebra, A Text/Workbook, Fourth Edition,* by Charles P. McKeague. It is designed for students and instructors to incorporate the use of graphics calculators into their algebra class.

The contents are organized into eight chapters. The Introduction and Chapter 1 give brief introductions to the CASIO, TI-81, and TI-82 graphic calculators. It is in these chapters that the basic operations of these calculators is discussed. Chapter 2 contains explanations on how to use graphics calculators to explore solving linear equations and inequalities. Chapter 3 makes use of graphs and programming to factor polynomials and solve polynomial and rational equations. Chapter 4 addresses equations of lines and inequalities. In chapter 5 a treatment of quadratic equations and parabolas is presented along with some of their applications. Chapter 6 shows how to exploit the graphic capabilities to visualize solutions to systems of linear equations. Chapter 7 contains discussions about functions, conic sections, and nonlinear systems. There is also included in chapter 7 a treatment of functions defined only by tables and how one may go about using the calculator to analyze these tables. Chapter 8 treats the exponential and logarithmic functions. At the end of each chapter there are investigations that ask the student to explore additional topics that are not sometimes considered in an algebra course. These range from order of operations to parametric representation of inverse functions. Each chapter includes explanation, worked examples and numerous exercises that require the use of a graphics calculator. Throughout this manual references are made to the text of Charles McKeague. For example, the reference: McKeague: Sections 7.1, 7.2, 7.3, 7.4; refers to McKeague's *Intermediate Algebra,* chapter 7, sections 1, 2, 3, and 4. The answers to all of the exercises are at the back of the manual.

This supplement is a compendium of our thoughts on how a graphics calculator could be used to enhance understanding in a course consisting of algebra. Some may say that a student in a traditional intermediate algebra course should not use calculators, especially those with graphics capabilities, but I say that if the few students who choose not to continue on with mathematics can be encouraged to do so by introducing them to a machine that will hopefully enhance understanding, then shouldn't we at least try it.

The intention here is to give a somewhat complete discussion of the details of how to operate graphics calculators and how to use them to explore the types of problems that arise in an algebra course. Many of the exercises here are routine and could be solved without such a sophisticated device, but in order to become proficient with the calculators we feel it is necessary to do some things which are routine. There are though many exercises that cannot be done except with some kind of graph being drawn and the interpreting of that graph. The graphics calculators make it possible to not just draw one or two graphs but several and with different scalings so that particular features of the graph can be easily discovered. As you work through this supplement you will find a great many programs that perform many tasks. It may seem that this will make things too easy, but how does an instructor encourage exploration without models to build on. So we invite others to take the programs here and explore their operation and thus provide a springboard to their own individual discoveries.

Some may wonder how the graphs were produced for this text. The graphs in the main body of the book were produced using *True BASIC® 2.5* for the Macintosh with the *Mathematician's Tool Kit® 1.01* from True BASIC Inc. The graphs in the Investigations at the chapter ends were ported using *TI-GRAPH LINK™* from Texas Instruments Inc.

James Angelos
Central Michigan University

Anita M. Johnston
Donal B. Staake, Jr.
Jackson Community College

Contents

Introduction **Calculator Overview** 1

Casio Overview ♦ TI-81/82 Overview

Chapter 1 **Basic Operations** 7

Screen Editing ♦ Arithmetic Operations
Using Memory ♦ Simple Programming
Graphing Functions

Chapter 2 **Linear Equations and Inequalities** 47

A Coordinate System ♦ Linear Equations
Inequalities ♦ Investigations

Chapter 3 **Polynomials and Factoring** 66

Polynomials ♦ Solving Equations
Synthetic Division ♦ Investigations: Rational
Exponents, Rational Expressions

Chapter 4 **Lines and Inequalities** 90

Lines ♦ Linear Inequalities
Investigations: Slope

Chapter 5 **Quadratic Equations** 102

Quadratic Equations ♦ Quadratic Inequalities
Graphing Parabolas ♦ Investigations: Quadratic
Inequalities

Chapter 6 **Systems of Linear Equations** 124

Systems of Two Equations ♦ Investigations:
Matrices

Chapter 7 **Functions and Conic Sections** 130

Explorations ♦ Tabular Functions
Inverse Functions ♦ The Circle
Ellipses and Hyperbolas ♦ Nonlinear Systems
Investigations: Inverses

Chapter 8 **Exponential and Logarithmic Functions** 164

Exponential Functions ♦ Logarithmic Functions
Investigations: Exponential Growth

Answers to Exercies 178

Index 209

Introduction

This manual is intended to acquaint you with graphics calculators and their use in precalculus mathematics. We will focus on the features of the CASIO family of calculators (the fx-7000G, fx-7500g, fx-7700G, fx-8000G, and fx-8500G) and the TI-81 from Texas Instruments that are useful in solving problems in algebra and trigonometry. You should be aware that this manual is not a substitute for the CASIO *Owner's Manual* or the TI-81 *Guidebook*. The basic features of the calculators will be examined here but not to the degree of detail found in the user manual accompanying your calculator. You do not need to know the entire user manual, but certainly enough to consult it for information when required.

CASIO Overview

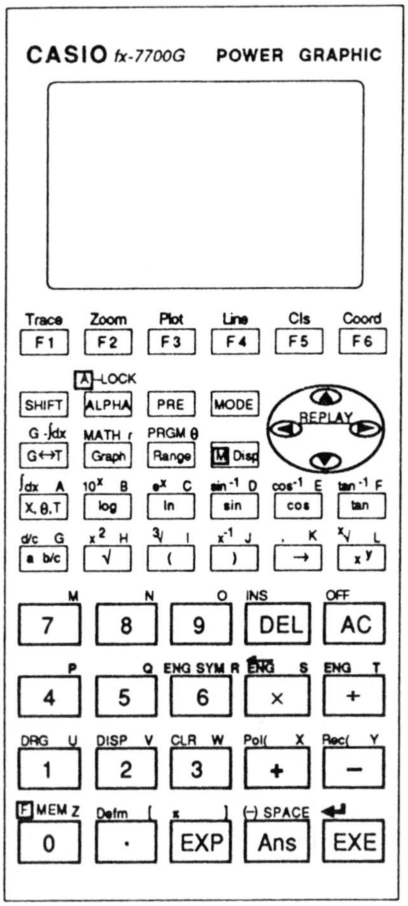

A reproduction of the CASIO fx-7700G is given in the figure to the left. The fx-8000G, fx-8500G and the fx-7000G have a different appearance. The fx-7500G is in a hinged hard shell case and has a different style of keyboard, but operates in the same manner as the others. The primary difference among the models is in the amount of memory available for storing programs. However, the fx-8000G and fx-8500G have file editors for named programs and optional printer and cassette recorder interfaces.

The first thing you might notice is the absence of an "=" key, which is found in most other brands of calculators. In its place at the lower right corner of the keyboard is the **EXE**, or *execute* key. Simply put, pressing this key will *execute* the command entered into the calculator, whether that be a simple arithmetic calculation or a graphics command.

Each of the 50 keys has multiple uses, the key itself and its other options obtained by using the **SHIFT** and **ALPHA** keys. The yellow **SHIFT** key accesses the key function written in yellow letters above and left the key. The red **ALPHA** key accesses the red letters or symbols above and right of a key.

For example, **SHIFT** √ accesses the "x^2" function while **ALPHA** √ accesses the letter **H**. Notice how the blinking cursor on the calculator changes as these buttons are pressed. To cancel one of these keys, simply press it again.

Let's now turn on the calculator. The **ON/OFF** switch is on the left side of the fx-7000G and fx-7500G and below and to the right of the screen on the fx-8000G and fx-8500G. The [AC] key is the **ON/OFF** switch for the fx-7700G. When the calculator is turned on you will see the *calculator status* screen. The fx-7000G, fx-7500G, fx-8000G, and fx-8500 screen is on the left and fx-7700G screen is on the right

```
**** MODE ****
sys mode  : RUN
cal mode  : COMP
   angle  : Rad
 display  : Norm

    Step    0
```

```
   RUN / COMP
G-type  : REC/CON
angle   : Rad
display : Nrm1
```

This screen gives information about the state of the calculator. If you are having difficulty seeing this screen, don't panic; the contrast of the screen can be adjusted to accommodate various lighting conditions. To change the contrast press [MODE] and hold down [⇒] to lighten the screen or press [MODE] and hold down [⇐] to darken the screen. The screen contrast on the fx-7700G is changed the same way except when then the [MODE] key is pressed a menu screen is shown. With this mode menu screen displayed the contrast is changed by holding down ◄, to lighten the screen or holding down ►, to darken the screen.

The first line of the status screen gives the **system** mode which should be the **RUN** mode for normal calculator functions. To put the calculator in this mode press [MODE] [1]. There are two other modes which are used for writing and deleting programs.

The second line (first line on the fx-7700G) tells the calculation mode. This should be the **COMP** mode, obtained by pressing [MODE] [+]. There are five (four on the fx-7700G) other calculation modes which are used when you deal with different number bases and statistical analysis and matrices. The second line of the fx-7700G gives the graph type and draw type. When the calculator is first turned on the graph type will be **REC** and the draw type will be **CON**. This means rectangular coordinate graphs where the points are connected by lines. There are three other graph types used for polar, parametric, and inequality graphs and there is one other draw type, **PLOT** where the points of the graph are not connected by lines. These are changed from a menu screen obtained by pressing [SHIFT] [MODE]. The third line gives the **angle** mode which can be either degrees, radians, or grads. For our purposes, radians will be used primarily. These are obtained by the following keystrokes.

	DEGREES	RADIANS
	[MODE] [4] [EXE]	[MODE] [5] [EXE]
fx-7700G	[SHIFT] [DRG] [F1]	[SHIFT] [DRG] [F2]

The fourth line is the **display** mode which tells the format that the calculator uses to display numbers on the screen. These are:

Norm Displays the full decimal precision of 10 digits. This format is set by the keystrokes [MODE] [9] [EXE] or [SHIFT] [DISP] [F3] for the fx-7700G.

Fix Sets the number of digits to the right of the decimal point to be displayed. For example, pressing [MODE] [7] [5] [EXE] , [SHIFT] [DISP] [F1] (**Fix**) [5] [EXE] on the fx-7700G, will cause the CASIO to display 5 digits to the right of the decimal point.

Sci Sets the number of digits to the right of the decimal point to be displayed, but using scientific notation. For example, pressing [MODE] [8] [5] [EXE] , [SHIFT] [DISP] [F3] (**Sci**) [5] [EXE] on the fx-7700G, will cause the CASIO to display 5 digits to the right of the decimal point in scientific notation.

All of the computations we will do in this manual will be performed in the **Norm** mode.

The fifth line tells how many steps are used in any calculation or sequence of commands. These steps are simply any character or single command that is entered on the screen. For example, if you enter **3 + 2 – 5** × **4**, the CASIO will count this as 7 steps. If at any time you wish to view the *calculator status* screen simply press and hold the key [M Disp].

The status screen is one of many that you will use to operate the calculator. The primary screen is the text screen, where all immediate calculations and graphics commands will be entered. To get to this screen from the status screen press [AC]. In the upper left corner of of the screen there will be a flashing line called the cursor. This marks where all characters will be entered and is capable of displaying up to 16 characters on a line and up to 8 lines of text. To clear the text screen press [AC].

The other screen that we will use is the graphics screen. It can viewed at any time by pressing [G↔T] (a key that toggles between the graphics and text screen) and is cleared by pressing [SHIFT] [Cls] , [SHIFT] [F5] (Cls) [EXE] on the fx-7700G. This screen has a resolution of 95 dots, which are called pixels, wide by 63 pixels high and we will learn to scale this screen to completely view virtually any graph of any function.

TI-81 Overview

A reproduction of the TI-81 is given in the figure below, the TI-82 is similar in appearance but with more features . Just as in the CASIO calculator there is no "=" key, in its place on the lower right side of the keyboard is the **ENTER** key. Simply put, pressing this key will *enter* or *execute* the command entered into the calculator, whether that be a simple arithmetic calculation or a graphics command.

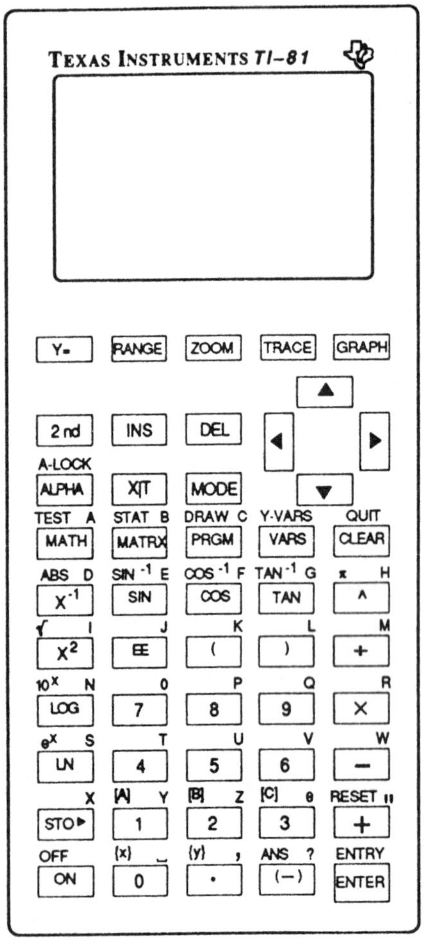

You will also notice that many of the keys have other options listed above them These options are accessed by using the **2nd** key and the **ALPHA** key. The blue **2nd** key accesses the key functions printed in blue above and to the left of the key. The **ALPHA** key accesses the letters and symbols above and to the right of the key. For example **2nd** $\mathbf{x^2}$ puts a square root (√) on the screen while **ALPHA** $\mathbf{x^2}$ puts the letter "**I**" on the screen. You will notice how the blinking cursor on the calculator screen changes as the **2nd** and **ALPHA** keys are pressed. To cancel any one of these keys, simply press it again.

Let's now turn on the calculator. The **ON/OFF** switch is a button on the bottom left hand side of the keyboard. To turn the calculator off press [2nd] [OFF]. When the TI-81 is turned on all that will be seen is a blank screen with the cursor flashing in the upper left hand corner of the screen. If you don't see the cursor there is no need to panic; all that you have to do is change the screen contrast. This is done by pressing [2nd] and holding down [▲] to darken the screen or [2nd] and holding down [▼] to lighten the screen. As you do this a number will appear in the upper right corner of the screen indicating the contrast setting. This screen is called the home screen and is one of the many that you will use to operate the calculator. It is where all immediate calculations are performed and is capable of displaying up to 16 characters on a line and up to 8 lines of text. The home screen is cleared by pressing CLEAR.

The TI-81 and TI-82 has a unique feature of putting most of its functions and commands in what are called menu screens. One of these screens contains all of the information about the calculator status and is given in the figure below. To display this screen press MODE. The items highlighted in black are the default settings of the calculator. In order to select any of these items, move the cursor with the arrow keys (◄ , ►, ▲, and ▼) to the item desired and press ENTER.

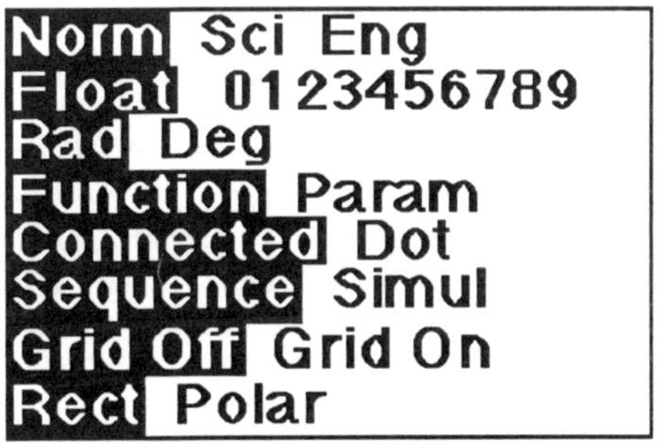

TI-81 Mode screen

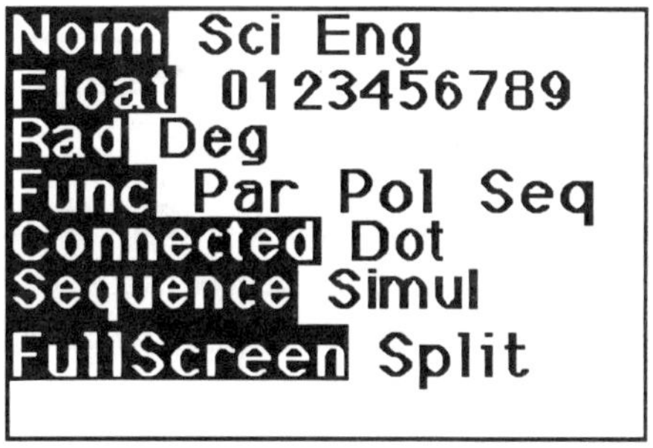

TI-82 Mode screen

The first two lines of this screen define the formats with which numbers are displayed. These formats are:

Norm Displays the full decimal precision of 10 digits. Uses scientific notation if a number is too large or too small.

Sci Puts all numbers in scientific notation.

Eng Puts all numbers in engineering notation. This is similar to scientific notation except the exponent on 10 is always a multiple of 3.

All of the computations in this book will be done in the **Norm** mode.

The line with the word **Float** sets the number of digits to the right of the decimal point to be displayed. For example, using the **Norm** and moving the cursor over the **5** and pressing ENTER will cause the calculator to display five digits to the right of the decimal point. To return to the full 10 digit display move the cursor over the **Float** and press ENTER.

The third line gives the angle mode which can be either degrees or radians. For our purposes radians will be primarily used.

The last five lines are modes that define types of graphs, how they are drawn, and how they are displayed. More will be said about these formats in Chapter 1.

After the desired modes are set, leaving the menu can be accomplished in any of three ways:

1. Pressing **2nd** **QUIT** returns to the home screen.

2. Pressing **CLEAR** returns to the previous screen.

3. Pressing another menu key.

The screen that we will make the most use of is the graphics screen which can be viewed at any time by pressing **GRAPH**. It is cleared by going to the **DRAW** menu screen (**2nd** **DRAW**) and selecting the first item in that menu, **ClrDraw**, which is done by moving the cursor over the number of the item and pressing **ENTER** twice. When this is done you will see on the home screen **ClrDraw** and the word **Done**. This indicates that the command has been executed properly.

Chapter 1

Basic Operations

Graphic calculators for all their sophistication are still useful for basic arithmetic calculations. The large screen display affords us the luxury of seeing an entire calculation written out before its evaluation. This feature alone would put the graphic calculator above all of the single number display calculators of the past. This chapter will guide you through the basic arithmetic operations of the CASIO, TI-81, and TI-82 graphic calculators. In addition we will explore the most powerful feature of these calculators, the ability to display graphical images of mathematical expressions. With a little practice you will learn to easily manipulate these graphs to solve many different kinds of problems. You should at all times while reading this manual have your calculator beside you and work the examples as you go. Finally, please explore and experiment on your own as well.

1.1 Screen Editing

As you perform various calculations you will notice that the cursor always is in the position where the next character will be entered and automatically moves to the next line when necessary. If you make an error and have to return to a particular character and correct it the keys [⇐], [⇒], [⇑], and [⇓] or (◄), (►), (▲), and (▼) on the CASIO and the keys [◄], [►], [▲], and [▼] on the TI-81/82 move the cursor left, right, up, and down respectively. To change a character move the cursor directly over the character and enter the new character.

There will be occasions when you need to insert or delete characters. The keys needed for these tasks are given below.

	CASIO	TI-81	TI-82
Insert characters directly to left of the cursor. Note that the cursor changes when in the insert mode.	[SHIFT] [INS]	[INS]	[2nd] [INS]
Delete the character under the cursor.	[DEL]	[DEL]	[DEL]

On the CASIO, notice the word **REPLAY** over the keys [⇐] or (◄) and [⇒] or (►). Pressing either of these keys returns to the screen the last command executed so it can be edited or reexecuted. Pressing [⇐] or (◄) returns the cursor to the end of the sequence and pressing [⇒] or (▶) returns the cursor to the beginning of the sequence. On the CASIO fx-8000G and fx-8500G you can also retrieve the last command executed by pressing [SHIFT] [ANS].

Caution: **If any key is pressed the last command becomes that key instruction. This will cause you to lose what was previously on the screen.**

To recall the last command entered on the TI-81/82 press [2nd] [ENTRY]. This can also be accomplished on the TI-81 when the [▲] key is pressed. The last command is also retained in memory even if the screen is cleared or if the calculator shuts off.

1.2 Arithmetic Operations

Both the CASIO and the TI-81 use standard algebraic order when evaluating arithmetic expressions. In any calculation, exponentiation is performed first, then multiplication and division, and then finally addition and subtraction. This ordering can be changed by grouping expressions within parentheses. Consider the following expression

$$7 - 8 \div 2 + 3 \times 1.7$$

This would be evaluated as

$$7 - 8 \div 2 + 3 \times 1.7 = 7 - 4 + 5.1$$
$$= 8.1$$

Using parentheses can change the value completely. For example,

$$(7 - 8) \div 2 + 3 \times 1.7 = -.5 + 5.1$$
$$= 4.6$$

Caution: **The negative sign and the subtraction sign are different keys. The [(-)] key changes the sign of a number while [−] is used for subtraction. These symbols are also displayed differently on the calculator screen. The negative sign is shorter than the subtraction sign.**

- **Example 1:** $5 - 4 \div 2 + 8 \times 3 = 27$

	Keystrokes	Display
CASIO	[5] [−] [4] [÷] [2] [+] [8] [×] [3] [EXE]	5−4÷2+8x3 27.

Keystrokes | Display

TI-81/82

[5] [−] [4] [÷] [2] [+] [8] [×] [3] [ENTER]

```
5-4/2+8*3
                27
█
```

- **<u>Example 2:</u>** 6 ÷ (-2) + 4 x 2.7 = 7.8

Keystrokes | Display

CASIO

[6] [÷] [(-)] [2] [+] [4] [×] [2] [·] [7] [EXE]

```
6÷-2+4x2.7
               7.8
```

TI-81/82

[6] [÷] [(-)] [2] [+] [4] [×] [2] [·] [7] [ENTER]

```
6/-2+4*2.7
               7.8
█
```

Notice that parentheses around the **-2** are unnecessary.

If an error occurs when entering a calculation both calculators produce warnings and allow you to move the cursor directly to the character that is incorrect. For example, suppose in keying in the calculation in example 1.2.2 above, the [−] was used instead of the [(-)] key. When the calculation is executed the calculators will respond with the following messages. The TI-82 message is similar.

```
6÷−2+4x2.7
   Syn ERROR
     Step 2
```

CASIO: Syntax Error Message

```
ERROR 06 SYNTAX
1: Goto Error
2: Quit
```

TI-81: Syntax Error Message

The CASIO tells you the step number prior to the error and you can move the cursor to the step where the error occurred by pressing either [⇐] (◄) or [⇒] (►). Note that on the the fx-7700G an error message will not be produced since it allows either [−] or [(-)] to be used for negation. However, when an error is made a syntax error message is displayed exactly as shown above.

The TI-81/82 gives you the option of going to the point where the error occurred or quitting the calculation completely. The last calculation sequence entered can be returned to the screen by pressing [▲] or [2nd] [ENTRY], with the cursor at the end of the sequence.

Any number of levels of parentheses can be used in a calculation. Both the CASIO and the TI-81/82 evaluate expressions with parentheses just as you would by hand, working from the inside out.

- <u>Example 3</u> -5 x (4 + 2 x (8 ÷ 4) – (-7 – 3 x (2 – 8))) = 15

Keystrokes | Display

CASIO

[(-)] [5] [×] [(] [4] [+] [2] [×] [(]
[8] [÷] [4] [)] [–] [(] [(-)] [7]
[–] [3] [×] [(] [2] [–] [8] [)]
[)] [)] [EXE]

```
-5x(4+2x(8÷4)-(-
7-3x(2-8)))
                15.
```

TI-81/82

[(-)] [5] [×] [(] [4] [+] [2] [×] [(]
[8] [÷] [4] [)] [–] [(] [(-)] [7]
[–] [3] [×] [(] [2] [–] [8] [)]
[)] [)] [ENTER]

```
-5*(4+2*(8/4)-(-
7-3*(2-8)))
                15.
```

Parentheses should be used whenever addition or subtraction is to be performed before a multiplication or division. For example if you intend to subtract 5 from 2 and divide by 6 then you should enter (2 – 5)÷6.

You will notice that each calculator has a key with the letters ANS. On the CASIO it is [Ans] and on the TI-81/82 it is [2nd] [ANS]. This key will access the last value calculated and can come in handy when doing a calculation with several levels of parentheses. By starting with the innermost expression, long calculations can be performed easily.

- <u>Example 4:</u> (8 – 10 x (2.7 – 6)) ÷ 3 = 13.66666667

Keystrokes | Display

CASIO

[2] [•] [7] [–] [6] [EXE]
[ANS] [×] [1] [0] [EXE]
[8] [–] [ANS] [EXE]
[ANS] [÷] [3] [EXE]

```
2.7-6
               -3.3
Ansx10
               -33.
8-Ans
                41.
Ans÷3
        13.66666667
```

Keystrokes | Display

TI-81/82

[2] [.] [7] [—] [6] [ENTER]
[2nd] [ANS] [×] [1] [0] [ENTER]
[8] [—] [2nd] [ANS] [ENTER]
[2nd] [ANS] [÷] [3] [ENTER]

```
                 -3.3
Ans*10
                  -33
8-Ans
                   41
Ans/3
          13.66666667
■
```

Computations using integral exponents require the use of three keys.

CASIO	[x^2] or [x^y] or [x^{-1}]
fx-7700G	[SHIFT] [x^2]
TI-81/82	[x^2] or [^] or [x^{-1}]

- **Example 5:** $2^5 \div 3^2 = 3.555555556 \left(= \frac{32}{9}\right)$

Keystrokes | Display

CASIO

[2] [x^y] [5] [÷] [3] [x^2] [EXE]

```
2xʸ5÷3²
          3.555555556
```

TI-81/82

[5] [÷] [3] [x^2] [ENTER]

[2] [^]

```
2^5/3
          3.555555556
■
```

- **Example 6:** $(2^{-3} \times 3^5) \div (2 \times 5^{-1}) = 75.9375$

Keystrokes | Display

CASIO

[2] [x^y] [(-)] [3] [×] [3] [x^y]
[5] [EXE]

```
2xʸ- 3x3xʸ5
               30.375
Ans÷(2x5 ⁻¹)
              75.9375
```

[ANS] [÷] [(] [2] [×] [5] [x⁻¹]
[)] [EXE]

TI-81/82

[2] [^] [(-)] [3] [×] [3] [^] [5] [ENTER]
[2nd] [ANS] [÷] [(] [2] [×] [5]
[x⁻¹] [)] [ENTER]

```
2^-3*3^5
                30.375
Ans/(2*5⁻¹)
               75.9375
█
```

• <u>Example 7:</u> $(-5 \div 4)^{-3} = -.512$

Keystrokes | Display

CASIO

[xʸ] [(-)] [3] [EXE]
[(] [(-)] [5] [÷] [4] [)]

```
(-5÷4) xʸ -3
                -0.512
```

TI-81/82

[(] [(-)] [5] [÷] [4] [)]
[^] [(-)] [3] [ENTER]

```
(-5/4)^-3
                 -.512
█
```

Both calculators have their own [√] keys to compute square roots. On the CASIO you simply press [√] followed by the number. The TI-81 requires pressing [2nd] [√]. Other roots are evaluated in the following ways. The CASIO will evaluate cube roots with the keys [SHIFT] [∛] and all other roots use the key [ˣ√]. The TI-81 has a cube root command in the [MATH] menu screen shown in the figure below right. The [MATH] of the TI-82 has both a cube root command and a nth root command.

```
MATH NUM HYP PRB
1:R▸P(
2:P▸R(
3: ³
4: ³√
5: !
6:°
7↓ r
```

TI-81 MATH menu

```
MATH NUM HYP PRB
1: ▸Frac
2: ▸Dec
3: ³
4: ³√
5: ˣ√
6: fMin(
7↓fMax(
```

TI-82 MATH menu

Other roots on the TI-81 are evaluated with the [^].

- **Example 8:** $\sqrt{1.44} = 1.2$

Keystrokes	Display
CASIO	
[√] [1] [.] [4] [4] [EXE]	√1.44 1.2
TI-81/82	
[2nd] [√] [1] [.] [4] [4] [ENTER]	√1.44 1.2 █

- **Example 9:** $\sqrt[3]{27} = 37.$

Keystrokes	Display
CASIO	
[SHIFT] [∛] [2] [7] [EXE]	³√27 −3.
TI-81/82	
[MATH] [▼] [▼] [▼] [ENTER] [2] [7] [ENTER]	³√27 −3 █

- **Example 10:** $\sqrt[5]{5/32} = 0.689864830.$

Keystrokes	Display
CASIO	
[5] [x√] [(] [5] [÷] [3] [2] [)] [EXE]	5ˣ√(5÷32) 0.6898648307
TI-81	
[(] [5] [÷] [3] [2] [)] [^] [5] [x⁻¹] [ENTER]	(5/32)^5⁻¹ .6898648307 █
TI-82	
[5] [MATH] [▼] [▼] [▼] [▼] [ENTER] [(] [5] [÷] [3] [2] [)] [ENTER]	5ˣ√(5÷32) 0.6898648307 █

- **Example 11:** $\sqrt[3]{4}/\sqrt{5} = 0.709907332.$

	Keystrokes	Display
CASIO	SHIFT, $\sqrt[3]{\ }$, 4, ÷, $\sqrt{\ }$, 5, EXE	$\sqrt[3]{}$4÷√5 0.709907332
TI-81/82	MATH, ▼, ▼, ▼, ENTER, 4, ÷, 2nd, $\sqrt{\ }$, 5, ENTER	$\sqrt[3]{}$4/√5 .709907332

- **Example 12:** $8^{5/3} = 32.$

	Keystrokes	Display
CASIO	8, x^y, (, 5, ÷, 3,), EXE	8x^y(5÷3) 32
TI-81/82	8, ^, (, 5, ÷, 3,), ENTER	8^(5/3) 32

- **Example 13:** $(\sqrt[4]{5} - 6) \div (\sqrt{3} + \sqrt[3]{8}) = -1.207017656.$

	Keystrokes	Display
CASIO	(, 4, $^x\sqrt{\ }$, 5, −, 6,), ÷, (, $\sqrt{\ }$, 3, +, SHIFT, $\sqrt[3]{\ }$, 8,), EXE	(4ˣ√5−6)÷(√3+³√8) -1.207017656
TI-81	(, 5, ^, 4, x^{-1}, −, 6,), ÷, (, 2nd, $\sqrt{\ }$, 3, +, MATH, ▼, ▼, ▼, ENTER, 8,), ENTER	(5^4⁻¹−6)/(√3+³√8) −1.207017656

TI-82

[( ] [4] [MATH] [▼] [▼] [▼] [▼] [ENTER] [5] [x⁻¹] [–] [6] [ )] [÷] [( ] [2nd] [√] [3] [+] [MATH] [▼] [▼] [▼] [ENTER] [8] [ )] [ENTER]

```
(5^4⁻¹-6)/(√3+³√8
)
          -1.207017656
```

1.2 Exercises

Evaluate each expression with your calculator

1. $2 + \frac{7}{5}$
2. $6 - 4x3 + \frac{16}{6}$
3. $(23.5 + 5x1.5)/5 - 11.2$
4. $4^3 + 7^2 - 3$
5. $\frac{3^4 5^{-2}}{\sqrt{2}}$
6. $\left(\frac{5}{\sqrt{2}}\right)^3$
7. $\frac{\sqrt{5}}{6 + \sqrt{3}}$
8. $(1 + \sqrt[3]{5})(1 - \sqrt[3]{5} + 5^{2/3})$
   (Try not putting the multiplication sign between the two expressions. What happens?)
9. $\sqrt{3^2 + 4^2}$
10. $\sqrt{(5-4)^2 + (-3-4)^2}$
11. $\sqrt[8]{231}$
12. $\frac{\sqrt[5]{-32}}{\sqrt{2}}$
13. $\sqrt{1/81}$
14. $5\sqrt[3]{3}\,(3^{2/3} - 5)$
15. $\frac{\sqrt{5.2} + \sqrt{1.1}}{\sqrt{5.2} - \sqrt{1.1}}$

1.3 Using Memory

The 26 letters **A-Z** that are accessed with the [ALPHA] key have as their primary use the storage of numbers. The following key sequences will store the number 2.5 in memory location **A** and recall the contents of memory location **A**.

CASIO	Storing	[2] [•] [5] [→] [ALPHA] [A]
	Recalling	[ALPHA] [A] [EXE]
TI-81	Storing	[2] [•] [5] [STO▸] [A] [ENTER]
	Recalling	[ALPHA] [A] [ENTER]

TI-82	Storing	[2] [•] [5] [STO▸] [ALPHA] [A] [ENTER]
	Recalling	[ALPHA] [A] [ENTER]

Notice that the TI-81 automatically activates the **ALPHA** keys when the [STO▸] key is pressed.

When a number is stored in a location it will remain there until you store a different value in its place. These stored numbers will remain in memory even when the calculator is shut off. The entire contents of the 26 memory locations on the CASIO can be cleared, i.e., set equal to zero by the keystrokes [SHIFT] [Mcl] [EXE], [SHIFT] [CLR] [F1] (Mcl) [EXE] on the fx-7700G. On the TI-81 there is only one way to reset the entire memory contents and that is by using the reset option. This is done by entering the key sequence [2nd] [RESET]. The following menu screen is displayed.

```
RESET
1: No
2:  Reset

STAT Bytes        0
PRGM  Bytes       0
Bytes  Avail   2400
```

This screen tells you how much memory is available for programming and statistical calculations. The calculator should be reset when you begin working with it and before you have entered any programs. On the TI-82, enter the key sequence [2nd] [MEM] and select the third item, **Reset** and press [ENTER]. The following screen is displayed.

```
RESET MEMORY
1:No
2:Reset


Resetting memory
erases all data
and programs.
```

To check how much memory is being used enter [2nd] [MEM] and select **Check RAM** and press [ENTER].

Caution: **If the calculator is reset after you have entered programs, they will be lost and have to be reentered.**

The memory location names **A-Z** can be used just as if they were numbers. They also behave like variables in algebraic expressions. For example, entering **AX** will perform multiplication on the values stored in the locations **A** and **X**. This allows you to evaluate expressions for several values of the variable.

• <u>Example 1:</u> Evaluate the expression $\mathbf{2A^2 - 1/A}$ for **A** equal to the values -5, 3, 0.01, 4, 117.

Keystrokes / Display

CASIO

[(-)] [5] [→] [ALPHA] [A] [EXE]
[2] [ALPHA] [A] [x²] [−] [1] [÷]
[ALPHA] [A] [EXE]

Display:
```
-5→A
                 -5
2A²-1÷A
               50.2
```

By repeating the above keystrokes for each value of **A** the following table is generated.

A	$2A^2 - 1/A$
-5	50.2
-3	18.333333333
0.01	-99.9998
4	31.75
117	27377.99145

TI-81

[(-)] [5] [STO▸] [A] [ENTER]
[2] [ALPHA] [A] [x²] [−] [1] [÷]
[ALPHA] [A] [ENTER]

Display:
```
-5→A
                 -5
2A²-1/A
               50.2
■
```

TI-82

[(-)] [5] [STO▸] [ALPHA] [A] [ENTER]
[2] [ALPHA] [A] [x²] [−] [1] [÷]
[ALPHA] [A] [ENTER]

Display:
```
-5→A
                 -5
2A²-1/A
               50.2
■
```

Repeating the above keystrokes for each value of A, the above table is generated.

You have probably noticed that this can be very tedious for complicated expressions. Let's investigate an easier way.

- **Example 1 (continued):**

CASIO

This calculator has two features which are very useful for evaluating expressions. The key sequence

[SHIFT] [?] [→] [ALPHA] [A] [EXE]

fx-7700G: [SHIFT] [PRGM] [F4] (?) [→] [ALPHA] [A] [EXE]

will prompt you with a **?** and wait until you input a value. When you input the value, it will be stored in the memory location **A**. Try it and check the value stored in **A**. Now key in the following sequence. (The **:** on the fx-7700G is obtained by press [SHIFT] [PRGM] [F6] (:)).

[SHIFT] [?] [→] [ALPHA] [A]
[:] [2] [ALPHA] [A] [x²] [−] [1] [÷]
[ALPHA] [A] [EXE]

```
?→A:2A²-1÷A
?
```

```
?→A:2A²-1÷A
?
-5
                50.2
?
-3
         18.33333333
```

The **?** prompts for a value of A. Input -5 and press [EXE]. Press [EXE] again. The **?** reappears, waiting for you to input the next value. Input -3 and press [EXE] Pressing [EXE] again begins the process over again. In this way you can evaluate the expression without reentering it each time the value of **A** is changed.

A word of explanation is in order. The [:] key is used to attach or concatenate commands. After you enter a value of **A** and press [EXE], the calculation $\mathbf{2A^2 - 1/A}$ is executed. When [EXE] is pressed again, the commands are reexecuted. The same thing can be accomplished without stringing commands together with the [:] key. Simply wherever there is a [:] enter instead [SHIFT] [↵]. This puts a "carriage return" at the end of each line and effectively attaches the lines together. This would change the keystrokes to the following

[SHIFT] [?] [→] [ALPHA] [A]
[SHIFT] [↵] [2] [ALPHA] [A] [x²]
[−] [1] [÷] [ALPHA] [A] [EXE]

```
?→A
2A²-1÷A
?
```

Caution: **Pressing any other key out of sequence will cause the CASIO to delete the commands, requiring you to reenter them.**

TI-81/2

The TI-81 has four storage locations that can be used to evaluate expressions. The TI-82 has 10 of these locations. Pressing the [Y=] causes the following Y variables list screen to be displayed.

```
:Y1 =█
:Y2 =
:Y3 =
:Y4 =
```

TI-81 Y variables screen

```
Y1 =█
Y2=
Y3=
Y4=
Y5=
Y6=
Y7=
Y8=
```

TI-82 Y variables screen

The cursor is poised at the first line and ready for you to enter the expression **$2A^2 - 1/A$**.

Keystrokes	Display
[2] [ALPHA] [A] [x^2] [—] [1] [÷] [ALPHA] [A] [ENTER]	:Y1 =2A²-1/A█

Notice the = sign is in inverted shading. This indicates that **Y1** is the active expression. To return to the home screen press [2nd] [QUIT], not [CLEAR]. Pressing [CLEAR] will erase the expression just entered. Now store -5 in memory location **A** ([(-)] [5] [STO▸] [A] [ENTER] or [(-)] [5] [STO▸] [ALPHA] [A] [ENTER]) and then key in [2nd] [Y-VARS]. On the TI-82 key in [2nd] [Y-VARS] [ENTER]. This activates the Y variables menu screen.

```
Y ON OFF
1:Y1
2:Y2
3:Y3
4:Y4
5:X1T
6:Y1T
7↓X2T
```

TI-81 Y-VARS screen

```
FUNCTION
1:Y1
2:Y2
3:Y3
4:Y4
5:Y5
6:Y6
7↓Y7
```

TI-82 Y-VARS screen

Select **Y1** by moving the cursor over the **1:** and press [ENTER]. This returns **Y1** to the home screen. Pressing **ENTER** again evaluates the expression stored in **Y1**, for **A** = -5, this value will be 50.2. To evaluate **2A^2 – 1/A** for the other values of **A**, repeat the following keystrokes for each value of **A**.

(-) **5** **STO▸** **A** **ENTER**
2nd **Y-VARS** (select **Y1**) **ENTER**
(**(-)** **5** **STO▸** **ALPHA** **A** **ENTER**
2nd **Y-VARS** **ENTER** **ENTER** on the TI-82.)

```
                 -3
Y1
               50.2
-3→A
                 -5
Y1
        18.33333333
■
```

The figure to the right is how the home screen may appear.

1.3 Exercises

Store the numbers in the corresponding memory locations and then evaluate the expression.

1. A = -0.5, A = 0, A = 3; $.5A^3 - A$
2. A = 2, B = -3; $\sqrt{A^2 + B^2}$
3. X = -4, X = -2, X = 0, X = 3/2; $2X^3 + 9X^2 - 2X - 24$
4. X = 1.5, Y = -1.5; $X^2 - 2XY + Y^2$
5. X = 2, H = -0.1, -0.01, -0.001, 0.001 0.01, 0.1; $\left(\sqrt{X + H} - \sqrt{X}\right)/H$

1.4 Simple Programming

1.4.1 Programming the CASIO

There are two system modes that you will use to enter, edit, and erase programs. Press **MODE** **2**. The screen display will appear something like this:

```
sys mode  :  WRT
cal mode  :  COMP
   angle  :  Rad
 display  :  Norm
  256 Bytes Free
 Prog 012_4567_9
```

fx-7000G, fx-7500G, fx-8000, fx-85000

```
   WRT / COMP

 4164 Bytes Free
P0 empty
P1 empty
P2 empty
P3 empty
```

fx-7700

Be certain that your calculator displays the same modes as the figure above. If you must make changes, perform the following while in the **RUN** mode (**MODE** **1**).

cal mode	MODE + (COMP)
angle mode	MODE 5 (SHIFT DRG F2 on the fx-7700G) (**Radians**)
display mode	MODE 9 (SHIFT DISP F3 on the fx-7700G) (**Norm**)

This screen tells you how many bytes are free and available for entering programs and which locations contain programs. There are 10 program storage locations numbered 0 through 9. The locations with a line instead of a number are the locations that contain programs. The fx-7700G has 38 program locations, **0-9**, **A-Z**, **r**, and θ. The locations that are empty are clearly marked.

Before keying in our first program we give a short synopsis of some of the keys we will use and their function.

Key	Keystroke	Position and Meaning
Quotes	ALPHA "	Found below the Prog key. Enclosing any character or word in quotes causes the calculator to display it.
fx-7700G	ALPHA F2	
Question Mark	SHIFT ?	Found above the → key. Used to prompt for numerical input. Programs will pause and wait for input.
fx-7700G	SHIFT PRGM F4	
Arrow	→	Used to assign a value to one of the memory locations **A-Z**, **r**, θ.
Display Marker	SHIFT ◢	Found above the : key. The ◢ character is placed at the end of a command line and causes the program to pause and display the value of the last calculation. When this is encountered the message – **Disp** – is displayed. To continue the program press EXE.
fx-7700G	SHIFT PRGM F5	
Carriage Return	SHIFT ↵	Found above the EXE key. This acts as a carriage return and is used to end a command line. Be careful not to press it twice at the end of a line; a syntax error

will result. No character is displayed to signify a carriage return.

Colon [:]
fx-7700G [SHIFT] [PRGM] [F6]

This is used to concatenate or string commands together in contiguous lines.

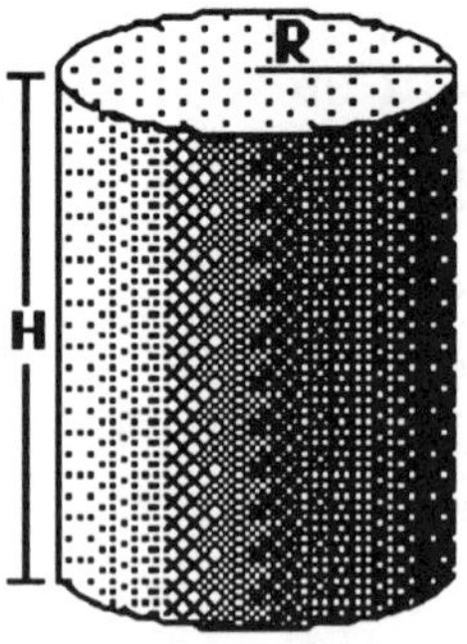

$V = \pi R^2 H$ (Volume)
$S = 2\pi R(R + H)$ (Surface Area)

Let's write a program that will find the volume and surface area of a right circular cylinder of radius **R** and height **H**. The figure to the left gives the appropriate formulas. In the program writing mode ([MODE] [2]) move the cursor with the keys [⇐] and [⇒] to program location 0 and press [EXE]. On the fx-7700G enter a number or letter and press [EXE]. You now have a blank screen with the cursor in the upper left corner of the screen. The keystrokes that follow will not have boxes around them. Each character will denote a keystroke. This program will be referred to as 1.4.1C. The "C" stands for CASIO.

Program 1.4.1C: Volume and Surface Area of a Cylinder

Keystrokes	Explanation
"R"?→R ↵	Displays the letter **R** and prompts for the input of the radius.
"H"?→H ↵	Displays the letter **H** and prompts for the input of the height.
"VOL" ↵	Displays the letters **VOL** which denotes volume.
πR²H→V ◢	Computes the the volume, stores the value in location **V** and displays the value.
"AREA" ↵	Displays the word **AREA**.
2πR(R+H)→S ◢	Computes the surface area, stores the value in location **S** and displays the value.

To run the program, return to the **RUN** mode ([MODE] [1]) and enter the sequence [Prog] [0] [EXE] ([SHIFT] [PRGM] [F3] [0] [EXE] on the fx-7700G). The letter **R** should appear with a **?**, prompting you to input a value for the radius. Input 3 for the radius and press [EXE]. The letter **H** appears with the **?** prompt again. Input 4 for the height and press [EXE]. The letters **VOL** are displayed followed by the value of the volume for a cylinder of radius 3 units and height 4 units. Pressing [EXE] once more displays the word **AREA** followed by the value of the surface area. The display will successively appear something like the two figures below.

```
Prog 0
R?
3
H?
4
VOL
          113.0973355
               - Disp -
```

```
3
H?
4
VOL
          113.0973355
AREA
          131.9468915
               - Disp -
```

Pressing [EXE] once more ends the program. You can tell this by the display; the value of the area is duplicated. If you press [EXE] again the program will restart. After the program is finished the values of the volume and surface area can be recalled from the memory locations **V** and **S**. If at any time you wish to stop the program press either [AC] or [MODE] [1].

Just as with the memory locations **A-Z**, programs are held in memory when the calculator is shut off. To erase a program you must enter the system mode **PCL** (program clear) by pressing [MODE] [3]. Move the cursor under the desired program location and press [AC]. On the fx-7700G enter the number or letter of the program and press [AC].

1.4.2 Programming the TI-81

The TI-81 and TI-82 are programmed by first accessing the **PRGM** menu screen. Press [PRGM]. The following menu screen will be displayed.

```
EXEC  EDIT  ERASE
1:Prgm 1
2:Prgm 2
3:Prgm 3
4:Prgm 4
5:Prgm 5
6:Prgm 6
7↓Prgm 7
```

TI-82 Program menu

```
EXEC  EDIT  NEW
1:Create New
```

TI-82 Program menu

With TI-81 working from this menu you can execute, edit, or erase as many as 37 programs on the . These are numbered 1, 2, 3, 4, 5, 6, 7, 8, 9, 0 or named by the letter A-Z and the

Greek letter θ (theta). Select the **EDIT** title and then a program location, **Prgm 1**, say. Press ENTER, and the edit screen for program 1 appears. This screen is shown below. With the TI-82, working from the program menu select and enter **NEW** and you are prompted to name the program. This screen is also shown below.

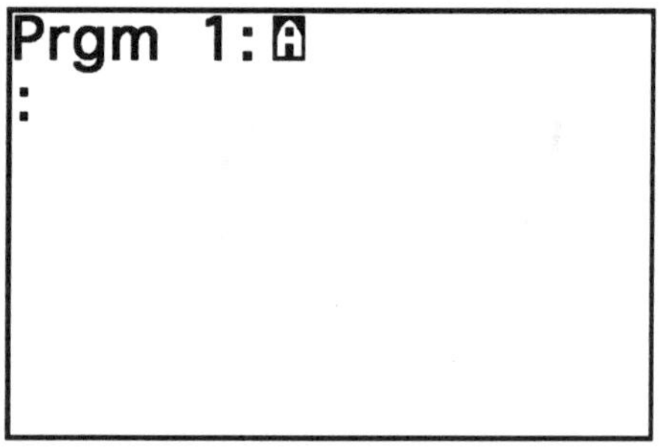

TI-81

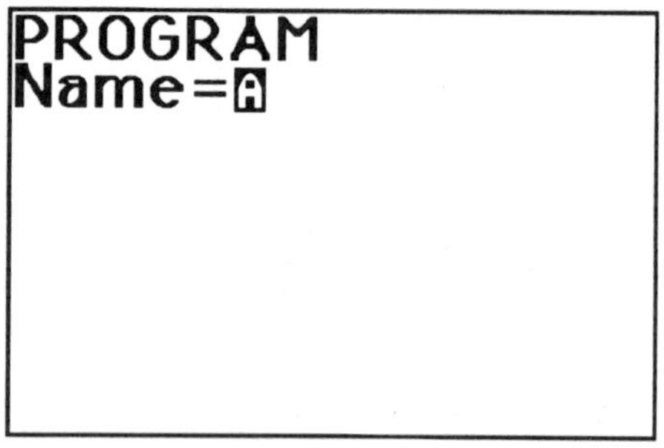

TI-82

The cursor is automatically in the **ALPHA** mode following the location name. You can now enter a name for the program, which can contain up to 8 characters. Before keying in our first program we will give a synopsis of the menu screens that contain the necessary programming commands and the keys that will be used to access these commands. (If you have a TI-82, enter the name **CYVLS** and press ENTER, you will now be able to view the control and imput/output menus.) While still in the edit screen for program 1, move the cursor to the second line by pressing ▼ once. Now press PRGM. There will appear the programming command menu screen. The two menus we will look at are the control (**CTL**) and the input/output (**I/O**) menus shown below.

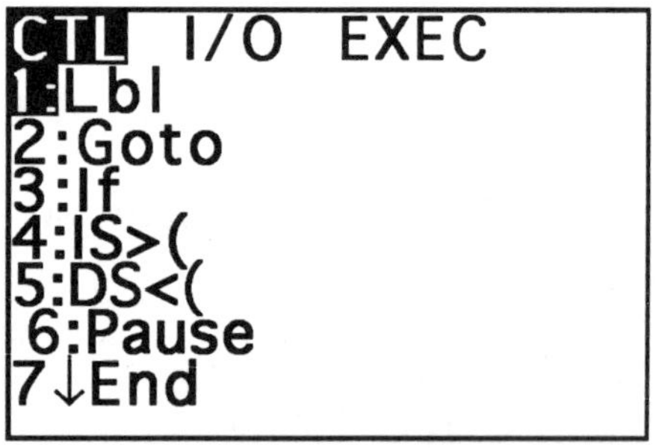

TI-81 Control Menu

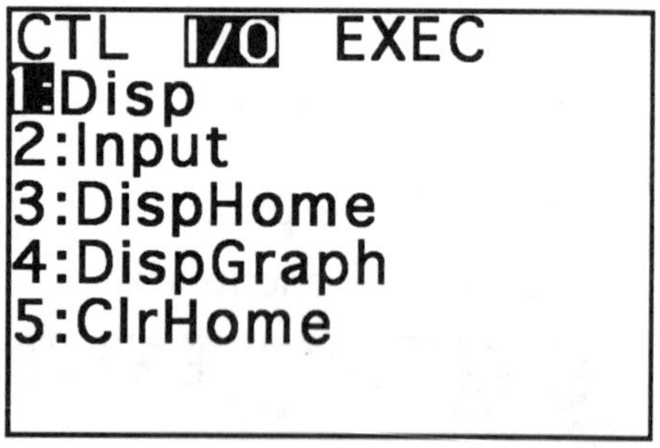

TI-81 I/O (Input/Output) Menu

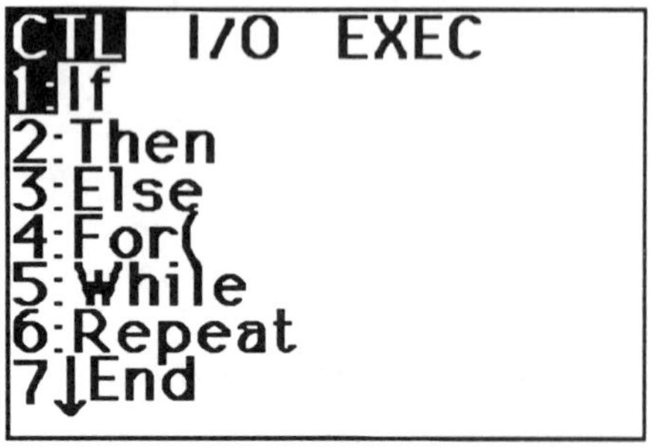

TI-81 Control Menu

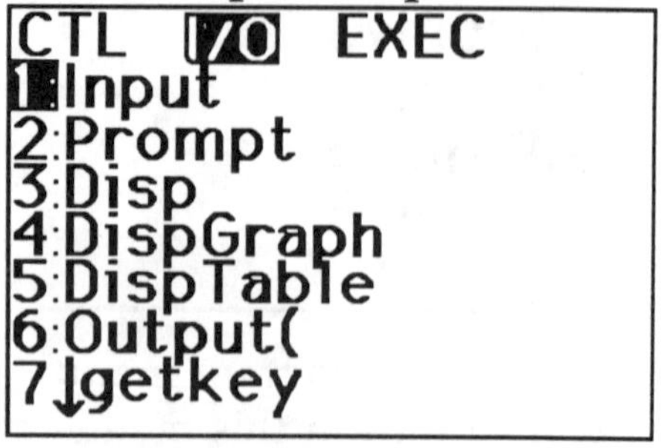

TI-81 I/O (Input/Output) Menu

The control menu contains the commands that are used to control program flow. The **I/O** menu holds the commands for displaying names, numbers and various screens. These commands are accessed by moving the cursor over the number of the command and pressing [ENTER].

Key or Command	Keystroke	Position and Meaning
Quotes	[ALPHA] ["]	Found above and right of the [+] key. Enclosing any word in quotes following the **Disp** command causes the calculator the display it.
Disp	[PRGM] [▶] [ENTER] (TI-82) [PRGM] [▶] [▼] [▼] [ENTER]	Found in the I/O menu when on line 2 or greater of the edit screen for a program. If followed by a quoted word or character, that word or character is displayed. If followed by an unquoted letter, the value stored in that letter is displayed.
Input	[PRGM] [▶] [▼] [ENTER] (TI-82) [PRGM] [▶] [ENTER]	Found in the I/O menu when on line 2 or greater of the edit screen of a program. Must be followed by a single letter. The program will pause, display a **?** and wait until a value is entered. To continue the program press [ENTER].
Arrow	[STO▶]	Used to assign a value to one of the memory locations.
Carriage Return	[ENTER]	Pressing this key when a command line of a program is ended begins a new line. No character is shown to signify a carriage return.

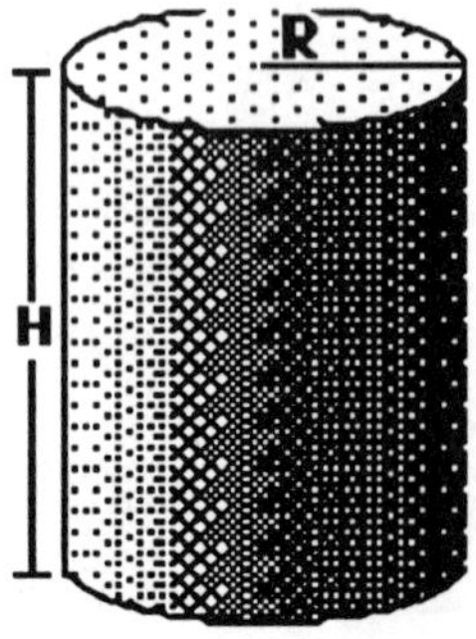

$V = \pi R^2 H$ **(Volume)**
$S = 2\pi R(R + H)$ **(Surface Area)**

Let's write a program that will find the volume and surface area of a right circular cylinder of radius **R** and height **H**. The figure to the left gives the appropriate formulas.

We will refer to this program as 1.4.1T. The "T" stands for TI-81. Begin by giving program 1 the name **CYLVS** for cylinder, volume, and surface area. You must begin entering commands on the second line of the edit screen. Each uppercase letter, or quote, or symbol is a single key stroke. Words that contain lowercase letters are obtained from the program command menus.

Program 1.4.1T: Volume and Surface Area of a Cylinder

Commands	Explanation
:Disp "R"	Displays the letter **R**.
:Input R	Prompts for the input of the radius.
:Disp "H"	Displays the letter **H**.
:Input H	Prompts for the input of the height.
:Disp "VOL"	Displays the letters **VOL**, which stand for volume.
:πR²H→V	Computes the volume and stores the value in location **V**.
:Disp V	Displays the value of the volume.
:Disp "AREA"	Displays the word **AREA**.
:2πR(R+H)→S	Computes the surface area and stores the value in location **S**.
:Disp S	Displays the value of the surface area.

Once you have entered the program correctly, return to the home screen by pressing [2nd] [QUIT]. To run the program press [PRGM], select the title **EXEC** and then the program number, which in this case is **1:**. Pressing [ENTER] returns the command **Prgm1** (**prgmCYLVS** on the TI-82) to the home screen. Pressing [ENTER] again will execute the program. The letter **R** should appear followed by a **?**, prompting you to input a value for the radius. Input 3 for the radius and press [ENTER]. The letter **H** appears with the **?** prompt again. Input 4 for the height and press [ENTER]. The letters **VOL** are displayed followed by the value of the volume for a cylinder of radius 3 units and height 4 units. This is followed by the word **AREA** followed by the value of the surface area. The display will successively appear something like the two figures below at the top of the next page.

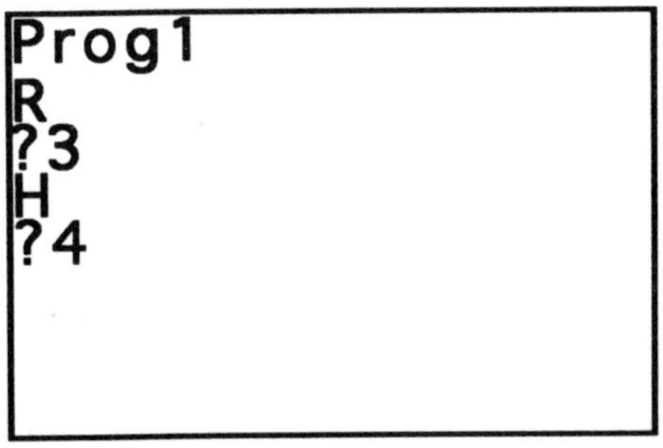

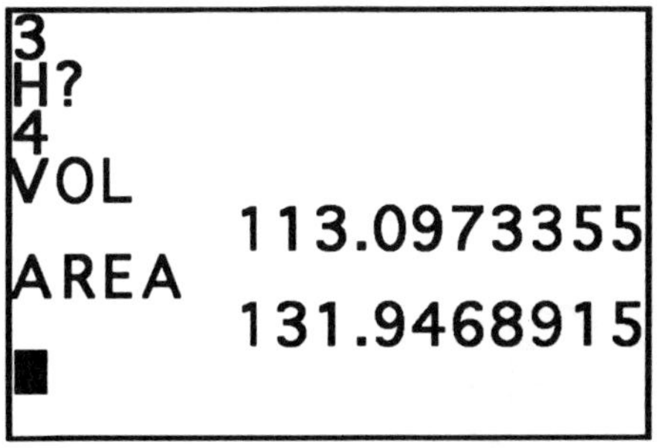

Pressing ENTER again will restart the program. After the program is finished the values of the volume and surface area can be recalled from their memory locations **V** and **S**. If during the execution of a program you wish to stop, press ON. This will produce the following screen. A similar screen is seen on the TI-82.

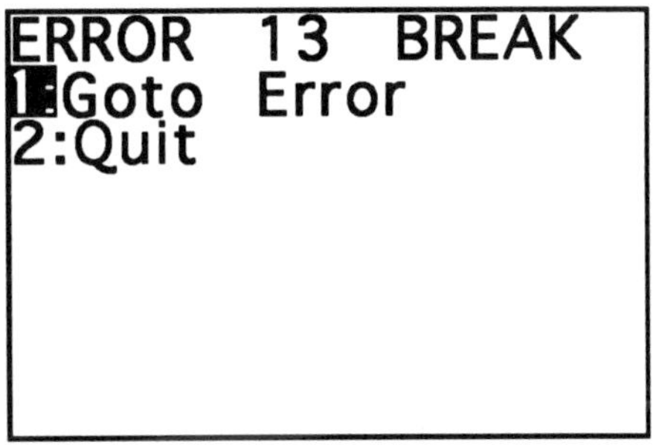

Select **2: Quit** and press enter. The word **Done** will appear on the screen (nothing on the TI-82 is displayed in this situation), signifying that the program was terminated. Just as with the memory locations **A-Z**, programs are retained in memory even when the calculator is shut off. To erase a program from memory press PRGM, select the title **ERASE**, then the program number or letter that you wish to erase and press ENTER. On the TI-82 enter 2nd MEM ▼ and then select and enter the 6th item Prgm. Select the program to be deleted and press ENTER.

1.4 Exercises

1. Use the program you just entered to find the volume and surface area of a cylinder with
   (a) **R** = 4, **H** = 3.
   (b) **R** = 1, **H** = 2.5.

2. Fixing the height of the cylinder to be 5, find the approximate value of the radius that produces a volume of 1000. [Hint: Start with R = 5; this gives a volume of 392.6990817. Now try R = 10, this gives a volume of 1570.796327, so the value of R we seek is between 5 and 10. Now try R = 7.5 = (5 + 10)/2. What value for the volume do you get? Now where is the value of R we seek? Continue in this fashion.]

3. Add to the program some lines that compute, store, and display the value of the quotient V/RS for values of $R = H = 1, 2, 3, 4, 5, 6, 7, 8, 9, 10$. Can you discover a joint relationship between **V**, **R**, and **S** when $R = H$? What about when $R = \frac{1}{2}H$?

4. Write and key in a program that computes the volume and surface area of a box of length **L**, width **W** and height **H**. Store the volume in location **V** and area in location **S**. Test the program by finding the volume of the box with following dimensions.
   (a) $L = 2, W = 5, H = 7$.
   (b) $L = 2.7, W = 1.2, H = 10$.

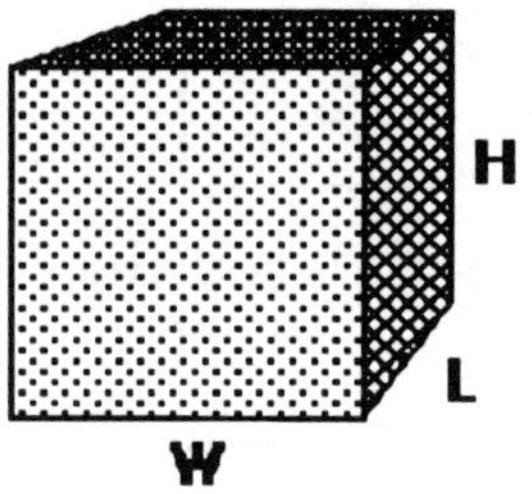

$V = LWH$
$S = 2LW + 2LH + 2WH$
$= 2(LW + LH + WH)$

5. Add to the program some lines that will compute, store and display the value quotient V/LS. Compute this quotient when the box is a cube, i.e., $L = W = H$ and when $L = W = H = 1, 2, 3, 4, 5, 6, 7, 8, 9, 10$. Can you discover a joint relationship between **V**, **L**, and **S** when the box is a cube? What about when $\frac{L}{H} = \frac{W}{H} = 2$? What about when $\frac{L}{H} = \frac{W}{H} = \frac{1}{2}$? Compare your answer with what you obtained exercise 3.

1.5 Graphing Functions

The feature that distinguishes both the CASIO and TI-81/82 and the one that first attracted the attention of so many students and teachers alike, and which convinced many of them to purchase one of these calculators, is the ability of the calculators to draw graphs of complicated functions. In the past many of us were taught to produce graphs by evaluating by hand or with a scientific calculator a function at a few points and then plotting the values on a coordinate system drawn on paper by hand. This was all fine and good for functions whose algebraic definition wasn't too complicated, but if the function was at all complicated it rendered this graphing method much too tedious and time consuming. This has all changed with the advent of the graphic calculator. We now have in our hands the capability of drawing the graph of almost any function that can be written down with a formula and as we shall see, some that have no formula. We will begin with a discussion of general ideas behind graphing on a calculator.

For any expression f(x), in order to produce its graph on the calculator screen we must first specify an interval in which the variable x will be. This interval is called in most cases the domain of the function f(x). This domain will take the form of $x_{min} \leq x \leq x_{max}$, where x_{min} and x_{max} are some real numbers satisfying $x_{min} < x_{max}$. The dependent variable, y, defined by the formula y = f(x) will then lie in some range which we will denote $y_{min} \leq y \leq y_{max}$. Unfortunately it is possible that for particular values of x the values of f(x) may be extremely large or f(x) may not exist. For example, if $f(x) = 1/(x-1)$ and x = 1.001, then $1/(1.001-1) = 1000$ and if x = 1, $1/(x-1)$ does not exist. So it may not be possible to capture an entire graph on the calculator screen. At best, we can only try to choose y_{min} and y_{max} so that enough of the graph of f(x) will be visible so that the features of interest can be viewed.

Each calculator has the facility for placing a coordinate system on the screen. The x variable specifies the horizontal range as $x_{min} < x < x_{max}$ and the vertical range as $y_{min} < y < y_{max}$. It also possible to set the scale of each coordinate axis, which is the distance between the tick marks. The following figure shows how the calculators will coordinatize the screen. This depicts the situation where $x_{min} < 0 < x_{max}$ and $y_{min} < 0 < y_{max}$ so the x- and y-

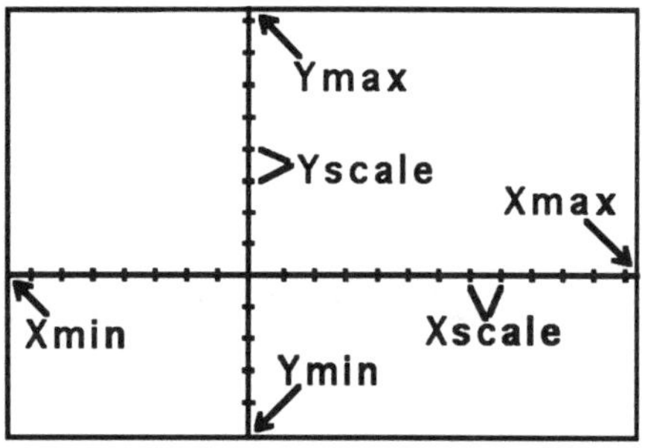

axes are visible. Throughout this manual, the screen ranges will be denoted by $[x_{min}, x_{max}] \times [y_{min}, y_{max}]$. For example, [-5, 3] x [-2, 8] indicates that the horizontal range is from -5 to 3 and the vertical range is from -2 to 8.

1.5.1 Graphing on the CASIO

The CASIO screen has a grid of dots called pixels. This grid is 63 pixels in height by 95 pixels wide. To set the screen range or to view the settings press **Range**. The cursor is under the value of **Xmin** and can be moved to any other value by using the keys **⇑** or **▲** and **⇓** or **▼**. To change any of the values simply enter the value and press **EXE**. The default setting for the screen range is given by [-4.7, 4.7] x [-3.1, 3.1] with both **Xscl** and **Yscl** equal to 1. This setting is obtained by pressing **SHIFT** **DEL** (**Range** **F1** on the fx-7700G). The default setting scales the screen in a 1:1 ratio, so graphs appear correctly scaled. In this range setting the pixels represent an increment of 0.1. In any other setting the increments are given approximately by

```
Range
Xmin:-4.7
 max:4.7
  sca:1
Ymin:-3.1
 max:3.1
  sca:1
```

$$\text{x increment: } \frac{\mathbf{Xmax - Xmin}}{95}$$

$$\text{y increment: } \frac{\mathbf{Ymax - Ymin}}{63}.$$

The key **G↔T** toggles between the graphic screen and the text screen. Let's begin graphing.

First clear the graphics screen by pressing **SHIFT** **Cls** **EXE** (**SHIFT** **F5** **EXE** on the fx-7700G) and set the range to the default setting by pressing **Range** **SHIFT** **DEL** (**Range** **F1** on the fx-7700G). Return to the text screen by pressing **Range** key once (twice on the fx-7700G). Notice the **Graph** key; press it and on the screen will appear **Graph Y =**. You can enter any single formula for a function using the variable **X**. Enter **ALPHA** **X**, press **EXE**, and the graph of y = x is drawn. You can now see the advantage of the default setting. The graph appears as it should, passing through the origin and making an angle of 45° with the x-axis.

The CASIO has several built-in functions that automatically set the range. These are functions that are graphed by simply entering **Graph** **(key)** **EXE**. Try entering **Graph** **x^2** **EXE** (**Graph** **SHIFT** **x^2** **EXE** on the fx-7700G). The function $y = x^2$ is drawn and the graphics screen looks like the figure at the top of the next page.

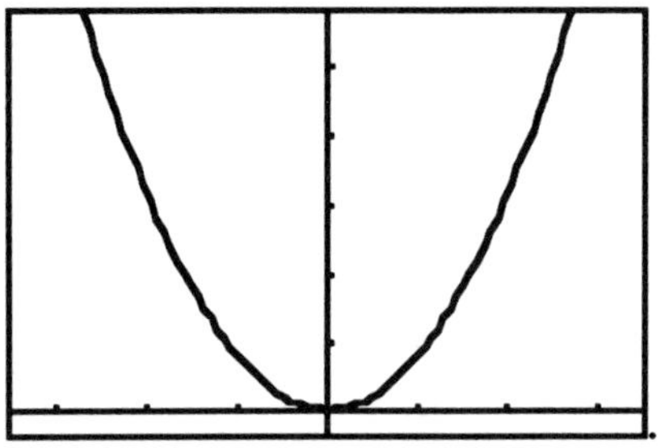

Check the range. You will see that it is [-7, 7] x [-2, 29]. Now try one of the other built-in functions; these include x^{-1},log x and 10^x among others. Once you have graphed each of these, check the range settings. Note also that the variable **X** need not be entered, when graphing one of these built in functions.

An important piece of information you need to get from a graph are values of the coordinates on the graph. This includes points where the function crosses the x-axis, called x-intercepts or zeros of the function. The CASIO has two features which facilitate this need: zooming and tracing.

• <u>Example 1:</u> Graph the function $y = 2x^2 + x - 3$ with the range setting [-2, 2] x [-6, 3], **Xscl** = **Yscl** = 1. Find the x-intercepts of the graph.

The graph will be similar to

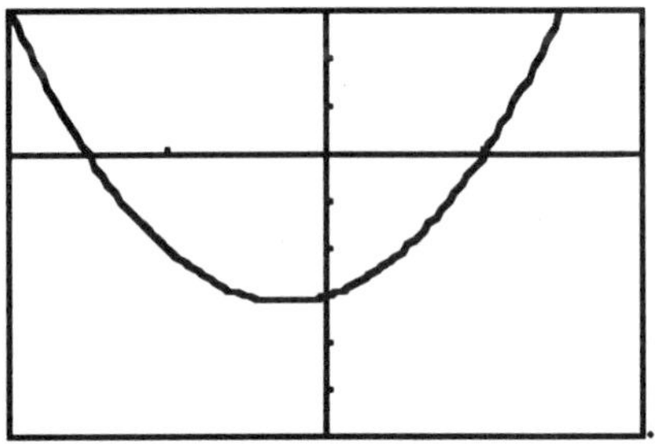

Press **SHIFT** **Trace** (**F1** on the fx-7700G). In the upper left corner of the screen their will be a flashing dot, a crosshairs on the fx-7700G, on the graph and the value of **X = -2** (**X=-2. Y=3.** on the fx-7700G) will appear across the bottom of the screen. Press and hold ⇒ or ►. What happens? This trace function reads off the x value on the curve and the dot is moved by pressing either ⇐ or ◄, or ⇒ or ►. To get the y value press **SHIFT** **X↔Y**. Now move the flashing dot to the first x-intercept. The approximate value you should get is **X = -1.489361702, Y = -0.05269514323**. On the fx-7700G if you press **F6** (**Coord**) a more accurate value of the point is displayed.

Because of the x- and y- increments that defined the screen resolution, this is the best you can do. To get a more accurate estimate of this x-intercept press [SHIFT] [×]. On the fx-7700G press [F2] (**Zoom**) followed by [F2] (**FCT**). Enter 2 for both **Xfct** and **Yfct**, then press [EXE]. Now press [F3] (**xf**). The graph is given below.

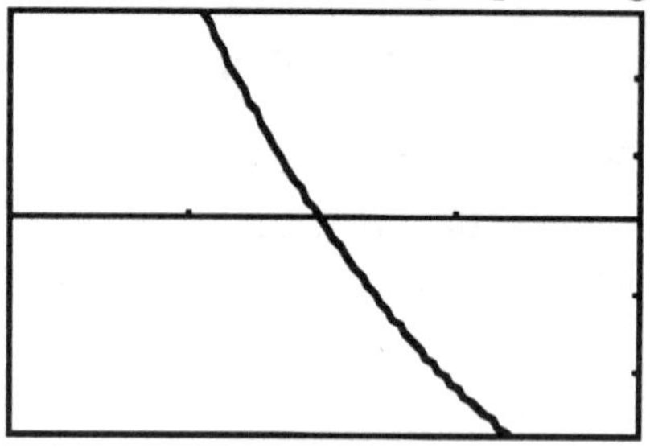

We have zoomed in on the point we traced to. Check the range. Using the trace function again move the dot to the intercept. We now have **X = -1.510638296, Y = 0.05341782668**. Zoom in again ([SHIFT] [×] ; [F2] [F3]) and trace to the intercept. The values are now **X = -1.500000002, Y = 1.E-08**. These will be **X = -1.5, Y = 0.** on the fx-7700G. It seems that this x-intercept is x = -1.5. Now with the dot on this intercept press [SHIFT] [÷], or [F2] (**Zoom**) [F4] (**x1/f**) on the fx-7700G. What happens? Check the range and compare it to the previous values. We have zoomed out on the point. Trace to the intercept again and zoom out again ([SHIFT] [÷], [F2] [F4]). The graph will now appear as

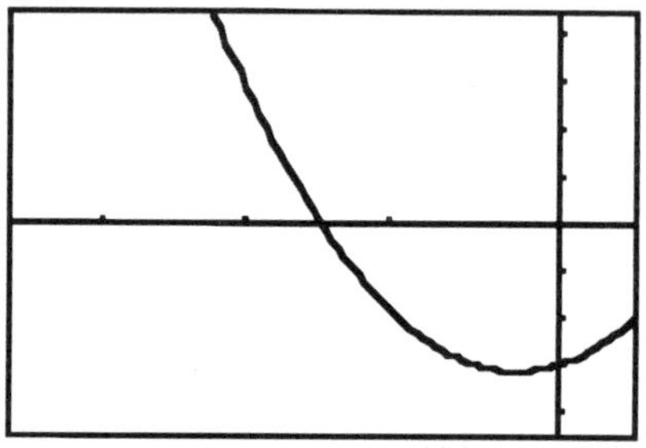

Trace to the far right point on the graph; this should be **X = -0.5, Y = -2**. Zoom out on this point, so we now have a more complete view of the graph. Trace to the right hand intercept and zoom in on it four times. The value you should get is x = 1.

There are other ways to zoom in or out. The easiest way is to change the range settings. The other way is to use the **Factor** function. (Not found on the fx-7700G.) To see how this function works, set the range to the default setting and key in [SHIFT] [Factor] [2] [EXE]. Check the range settings. What has happened? Now enter [SHIFT] [Factor] [·] [5] [EXE] and check the range settings. Using this **Factor** function we are able to zoom in or out to a lesser or greater degree than by using the automatic zoom. However, this feature is much easier to use if we place the graphing commands in a program.

• <u>Example 2:</u> Create a program that graphs the expressions $y = 1 - x^2/4$ and $y = 5x^3 - 3x$ and find the points of intersection of the two graphs.

Put the following program in location 2.

Program 1.5.1C: Function Graphs

Keystrokes	Explanation
Graph Y=5Xx^y3–3X ↵	Draws the graph of $y = 5x^3 - 3x$.
Graph Y=1–.25X² ↵	Draws the graph of $y = 1 - x^2/4$.

After keying in the program return to the **RUN** mode ([MODE] [1]) and set the range to the default settings. Execute the program by entering [Prog] [2] [EXE] or [SHIFT] [PRGM] [F3] (**Prg**) [2] [EXE]. The graphs will appear something like the graph below.

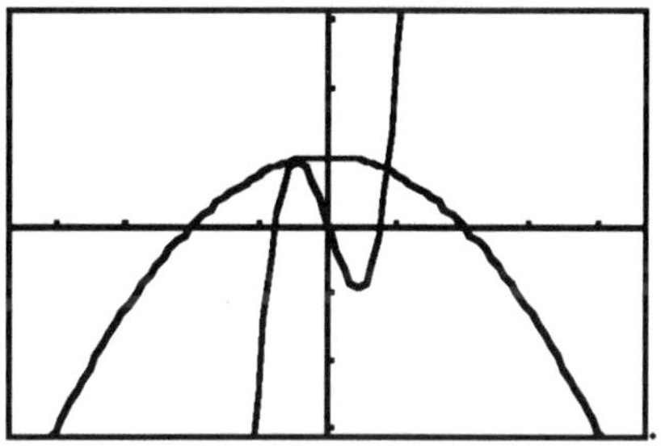

Using the trace, move the dot to the first point that appears to be an intersection point. This is about **X = -0.5**. Note that the trace works on the last graph drawn however, on the fx-7700G you can switch the trace to the other graph by pressing the ▲ key or the ▼ key. Return to the text screen, enter [SHIFT] [Factor] [5] [EXE]. For the fx-7700G press [F2] twice and set the values of **Xfct** and **Yfct** equal to 5 and press [EXE] followed by [F3]. Reexecute the program. (Not necessary on the fx-7700G.) The screen shows

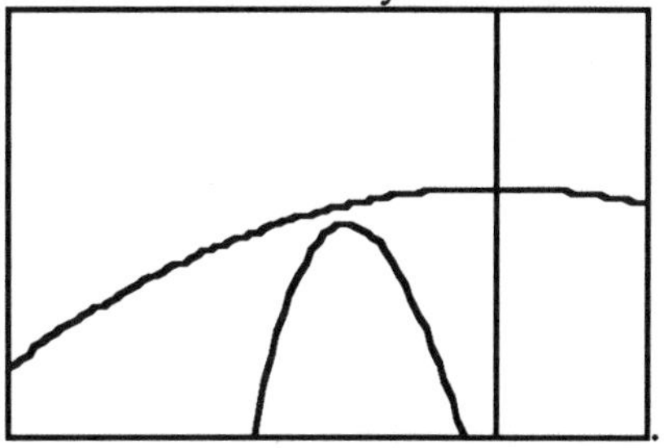

This tells that the graphs do not intersect at this point. Now reset the range to the default setting and run the graphing program again. Trace to the other intersection point, about **X = 0.9** and as before use the **Factor** function with a value of 10 (**Xfct** and **Yfct** equal to 10 on the fx-7700G). The graph now look like the figure at the top of the next page.

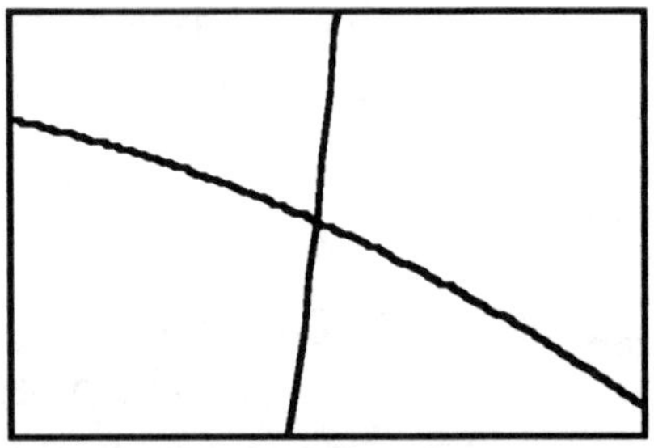

Again trace to the intersection point (**X = 0.88**) and zoom in again by a factor of 10. This time when you trace to the intersection point you should have the values **X = 0.884, Y = 0.804636**. Checking the range you will find **Xmin = 0.833, Xmax = 0.927, Ymin = 0.779,** and **Ymax = 0.841**. The differences **Xmax – Xmin** and **Ymax – Ymin** give a measure of the accuracy of the estimate for the intersection point. Tracing to the point and then using **Factor** with a value of 10, (**Xfct** and **Yfct** equal to 10 on the fx-7700G) 4 more times gives the estimate for the intersection point **X = 0.88428366, Y = 0.8045106022**. The range settings indicate that this value is accurate to approximately 6 decimal places.

The fx-7700G also has in the **Zoom** menu the command **BOX**. This can be used to box in a region of the screen. Return to the default range and execute the program given in this example. After the graphs are drawn press [**F2**] followed by [**F1**] (**BOX**). You will see a flashing dot, the center of a cross hairs, at the center of the screen. Using the arrow keys move the cross hairs to the point indicated in the figure below left and press [**EXE**]. Now move the dot to the point indicated in the figure below right. As you do so

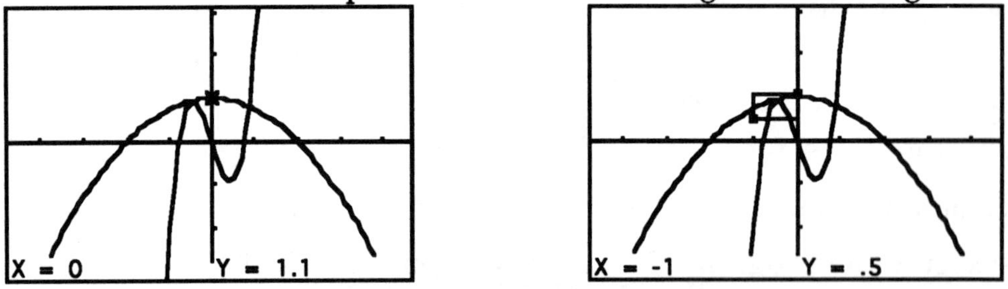

a box will be drawn enclosing the point of interest. Once you have enclosed the point press [**EXE**] and the box is blown up to the size of the screen, thus zooming in on the point. The following figure is produced revealing that the graphs do not intersect in the second quadrant.

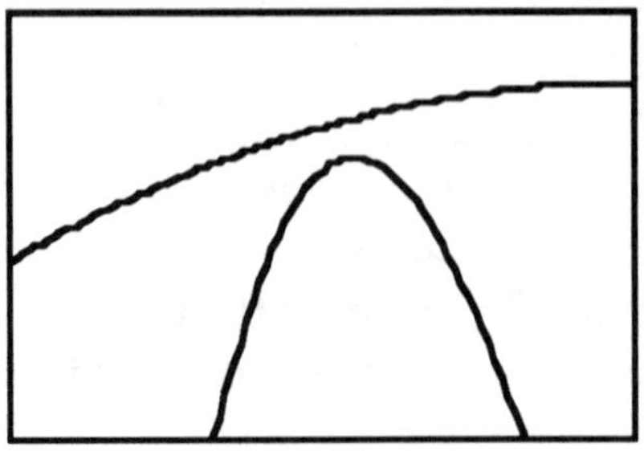

Caution: **Care must be used when using the Factor function. If the flashing dot isn't very close to the point of interest and the factor value is large, it is possible to lose part or all of the graph. If this happens reset the range by hand and start over.**

CASIO Graphing Review

Function	Keystrokes	Explanation
Range	[Range]	Used to the set the screen coordinates **Xmin**, **Xmax**, **Xscl**, **Ymin**, **Ymax**, and **Yscl**. Pressing [SHIFT] [DEL] enters the default range settings [-4.7, 4.7] x [-3.1, 3.1] with **Xscl** = **Yscl** = 1. [Range] [F1] on the fx-7700G.
Clearing the Graphics Screen	[SHIFT] [Cls] [EXE] fx-7700G [SHIFT] [F5] (Cls) [EXE]	Clears the graphics screen and readies it for another graph.
Graph	[Graph]	Puts the expression **Graph Y=** on the screen. Enter any single formula for a function using the variable **X** only. Pressing [EXE] starts the graphing.
Trace fx-7700G	[SHIFT] [Trace] [F1] (Trace)	Gives the values of the coordinates on the graph as per position of the flashing dot or crosshairs. The dot is moved by the keys [⇐] and [⇒], ◄ and ► on the fx-7700G.
Automatic Zoom in	[SHIFT] [X] fx-7700G [F2] (Zoom) [F3] (xf)	Zooms in on any point that is traced to. It does this by redrawing the graph with the range settings halved. On the fx-7700G, range settings are divided by the values of **Xfct** and **Yfct**.

Boxing in

fx-7700G [F2] (Zoom) [F1] (BOX) Used to draw a box around a point and blow that up to the size of the screen.

Automatic Zoom out

[SHIFT] [÷] Zooms out on any point that is traced to. It

[F2] (Zoom) [F4] (x1/f) does this by redrawing the graph with the range settings doubled. On the fx-7700G, range settings are multiplied by the values of **Xfct** and **Yfct**.

Set Zoom

Factors [SHIFT] [Factor] (number) [EXE] Sets the factor by which the range settings are divided. The value of **number** must be

or

[SHIFT] [Factor] (number) positive. If **number** > 1, you zoom in and if

[SHIFT] [,] (number) [EXE] **number** < 1 you zoom out. The second set of keystrokes is used to set different factor values of **X** and **Y**. Set from the **Zoom** [F2] menu under [F2] (**FCT**) on the fx-7700G.

1.5.1 Graphing on the TI-81/82

The TI-81 screen has a grid of dots called pixels which is 64 pixels in height by 96 pixels wide. The TI-82 screen is 63 pixels in height by 95 pixels wide. To set the screen range or to view the settings press [RANGE], [WINDOW] on the TI-82. The TI-82 has the cursor on the first line which defines **Xmin** and can be moved to any other line by using the keys [▲] [▼]. To change any of the values simply key in the number and press [ENTER]. When [WINDOW] is pressed on the TI-82 press [ENTER] or [▼] to move to the line defining Xmin. Pressing [▲] or [▼] moves to any other line. To change any of the values simply key in the number and press [ENTER].

```
RANGE
Xmin=-4.8
Xmax=4.7
Xscl=1
Ymin=-3.2
Ymax=3.1
Yscl=1
Xres=1
```

This then moves the cursor to the next line. The last line of this screen is the value of **Xres** (not on the TI-82) which must be an integer between 1 and 8. The calculator takes this value and divides it into 96; the quotient is the number of points at which the function is be evaluated and the number of pixels that will be turned on when it is graphed. Higher values of **Xres** produce graphs faster but can look like broken line segments. Pressing [CLEAR] will delete a value from a line and pressing [2nd] [QUIT] returns to the home screen.

A useful setting that scales the screen in a 1:1 ratio are:

TI-81: [-4.8, 4.7] x [-3.2, 3.1] with **Xscl** = 1, **Yscl** = 1, and **Xres** = 1
TI-81: [-4.7,4.7] x [-3.1, 3.1] wigh **Xscl** = 1, **Yscl** = 1.

We will call these the default settings. In this setting the pixels represent an increment of 0.1. In other settings the increments are given approximately by

x increment: TI-81: $\frac{\text{Xmax} - \text{Xmin}}{96}$ TI-82: $\frac{\text{Xmax} - \text{Xmin}}{94}$

y increment: TI-81: $\frac{\text{Ymax} - \text{Ymin}}{64}$ TI-82: $\frac{\text{Ymax} - \text{Ymin}}{62}$.

The graphics screen is cleared by accessing the **ClrDraw** command from the **DRAW** menu. Here are the keystrokes: [2nd] [DRAW] [ENTER] [ENTER]. The graphics screen can be viewed at any time by pressing [GRAPH]. This will redraw the graphs if any changes were made to the functions or simply put the graphs back on the screen. Let's begin.

First clear the graphics screen and set the range to the default. Pressing [ZOOM] and selecting and entering **ZDecimal** will set the TI-82 to the default range. Go to the Y variables list by pressing [Y=]. Here you will enter expressions defining functions using the variable **X** only. On the **Y1=** line enter the variable **X**. The screen will appear as

```
:Y1=X
:Y2=
:Y3=
:Y4=
```

TI-81 Y variables screen

```
Y1=X
Y2=
Y3=
Y4=
Y5=
Y6=
Y7=
Y8=
```

TI-82 Y variables screen

If there are other expressions entered for **Y2**, **Y3**, or **Y4**, etc., turn off these expressions by moving the cursor over the "=" sign and pressing [ENTER]. Recall an expression is active if the = looks like ■. After entering **X** and turning off the all other expressions, press

GRAPH. The graph of y = x is then drawn. You now see the advantage of this default range setting. The graph appears as it should passing through the origin and making an angle of 45° with the x-axis. Now enter **X²** for **Y2** and press **GRAPH**. The graphs of y = x and $y = x^2$ are now drawn with the screen looking like this

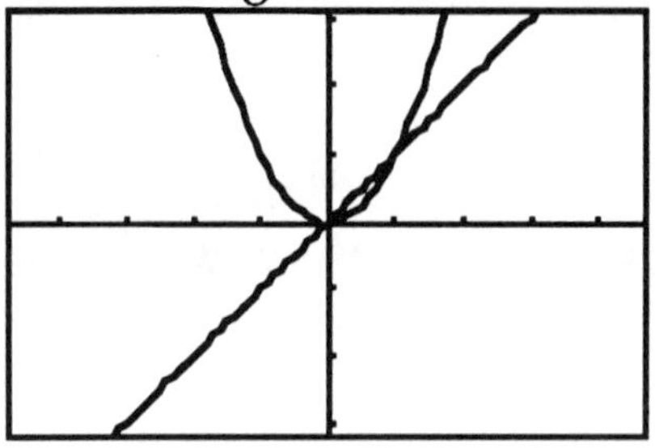

Now is a good time to return to the **MODE** screen and look at the other settings. Press **MODE**. The last 5 settings are for the presentation of graphs. **Function** tells the calculator we are drawing expressions that are given as a single function of x as opposed to parametric equations. **Connected** will turn on pixels defined by y = f(x) and then draw a line between the pixels whereas **Dot** simply turns on the pixels and does not draw a line between them. Select **Dot** with the flashing cursor and press **ENTER**. **Sequence** and **Simul** draw graphs one at a time and simultaneously, respectively. Select and enter **Simul**. **Grid On** will cause pixels that are defined by **Xscl** and **Yscl** to be turned on over the entire graphics screen. Select and enter **Grid On** and then press **GRAPH**. You should see the following screen. On the TI-82 selecting **Dot** and **Simul** and in the **WINDOW** menu, under the title **FORMAT**, selecting **GridOn** will produce the same screen.

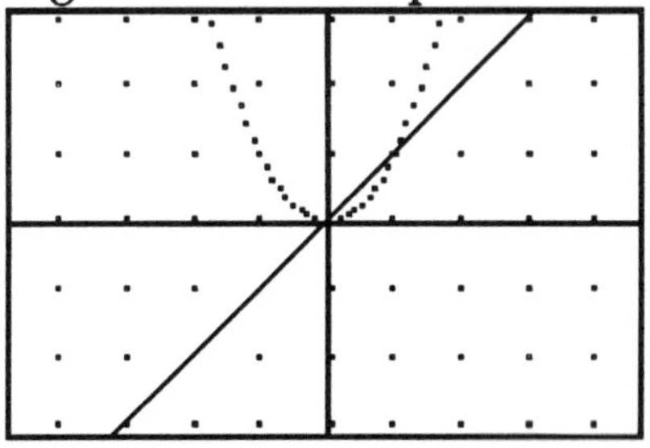

An important piece of information you need to get from a graph are values of the coordinates on the graph. This includes points where the function crosses the x-axis, called x-intercepts or zeros of the function. The TI-81/82 has two keys which make this process simple: **ZOOM** and **TRACE**.

• <u>Example 3:</u> Graph the function $y = 2x^2 + x - 3$ with the range setting [-2, 2] x [-6, 3], and **Xscl** = **Yscl** = 1. Find the x-intercepts of the graph.

Enter in the **Y1** variable the function **2X² + X – 3** and set the range to be [-2, 2] x [-6, 3]. Be certain that all other expressions in the Y variable list are turned off. Set the modes to be **Connected, Simul**, and **Grid Off**. Draw the graph. You should see something like the

the following figure.

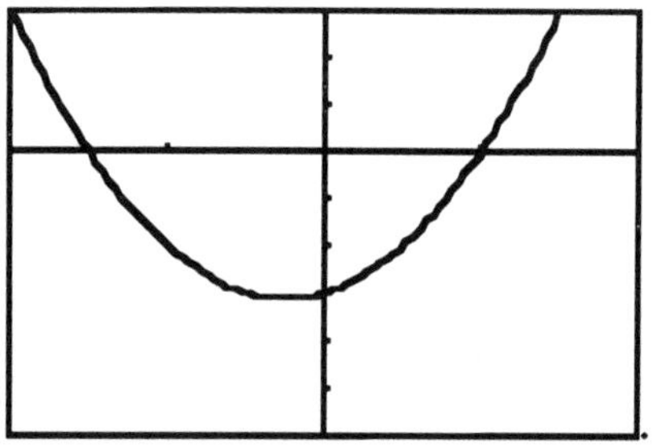

Press **TRACE**. On the y-axis where the graph crosses there will be a flashing box with an "**X**". At the bottom of the screen will be the x and y coordinates of the flashing box. Press and hold either ◀ or ▶. What happens? Move the flashing box to the left most x-intercept. The approximate value is **X = -1.494737, Y = -.0262604**. Slightly different values will be seen on the TI-82. In the **ZOOM** menu, the fourth item is **Set Factors.** Select and enter **Set Factors**. Enter for both **XFact** and **YFact** the value of 2. On the TI-82 select the **MEMORY** title from the **ZOOM** menu. Select the 4th item, (**SetFactors...**). Set both **Xfact** and **YFact** to 2. We can now obtain a more accurate estimate by using the functions in the **ZOOM** menu. Press **ZOOM**, select **Zoom In**, and press **ENTER**. The graph will reappear with a flashing dot. This dot can be moved to any place on the screen with the arrow keys, but for now move it to the left most x-intercept. Press **ENTER** again and the graph is redrawn, but now we have zoomed in on the intercept. See the figure below. Slightly different values will be seen on the TI-82. Press **TRACE** again, move to

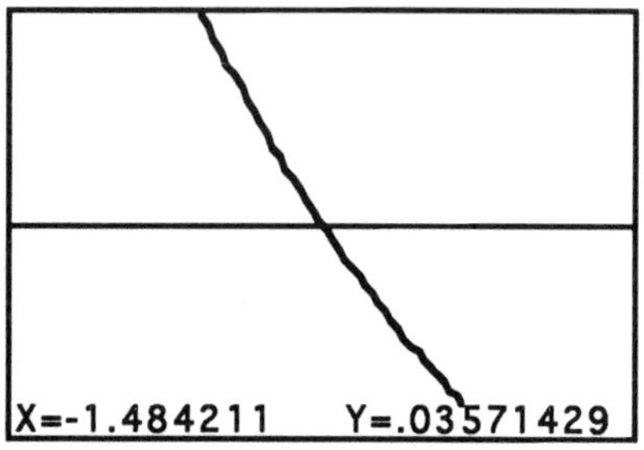

the intercept and again select **Zoom In** from the **ZOOM** menu and press **ENTER** twice. Trace to the point again and we now have **X = -1.5, y = 5.25E-10 = 5.25** $\mathbf{x10^{-10}}$. (On the TI-81 his value is gotten by moving slightly above what appears to be the actual intercept. The reason for this is with an even number of pixels in each direction there is no pixel at the exact center of the screen.) We know from the original graph of the function that there is another x-intercept on the positive side of the x-axis, so pressing **TRACE** and holding down ▶ the flashing box will eventually go off the screen and then the part of the graph that is visible will scroll slowly to the left and off the screen. Continue tracing anyway and soon the y-axis will appear, followed eventually by the other intercept. Tracing all the way to this other intercept, we get an estimate of **X = .9947364, Y = -0.262604**. The TI-82 will read **X = 1, Y = 0**. If you now repeat the zooming in procedure 4 times you obtain the estimate **X = 1.0003289, Y = .00164494**. It appears the other intercept is x = 1.

The TI-81 has another way to zoom in on a point; this is the **Box/ZBox** function in the **ZOOM** menu.

• <u>Example 4:</u> Graph the functions $y = 1 - x^2/4$ and $y = 5x^3 - 3x$ using the default range setting. Find the points of intersection of the two graphs.

Enter **1–.25X²** for **Y1** and **5X^3–3X** for **Y2**. Set the range to be TI-81: [-4.8, 4.7] x [-3.2, 3.1], TI-82: [-4.7, 4.7] x [-3.7, 3.1 with **Xscl** = **Yscl** = 1 and draw the graphs, which will look like

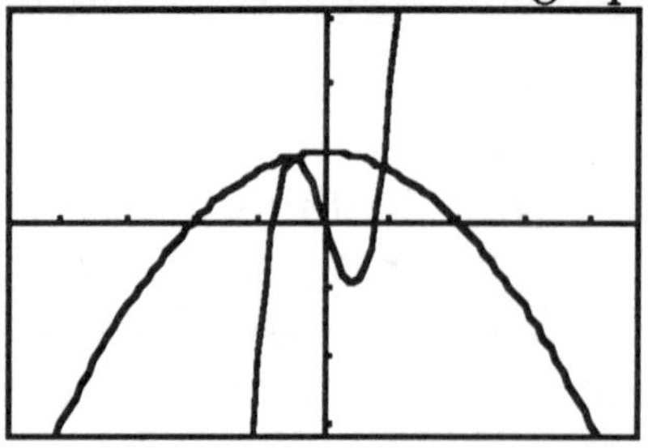

There appear to be two points of intersection. Before we zoom in on this point we note in passing that when the trace is active you can move the flashing box from one curve to the other by pressing [▲] or [▼]. Now let's zoom in on the apparent intersection point in the second quadrant. From the **ZOOM** menu select and enter **Box/ZBox**. On the graph there will be a flashing dot at the screen center. Using the arrow keys move it to the point indicated in the figure below left and press [ENTER]. The dot will change to a small box.

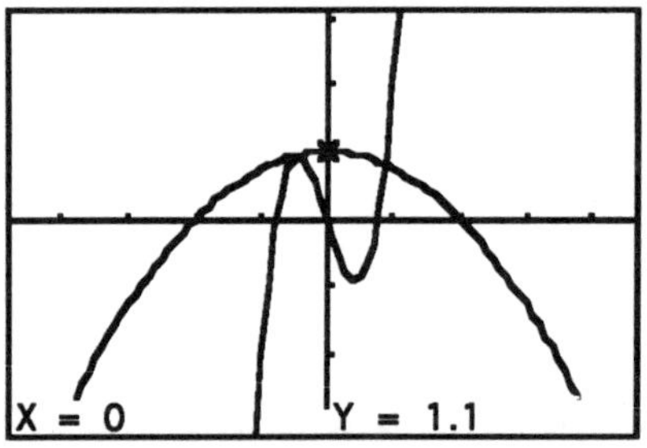

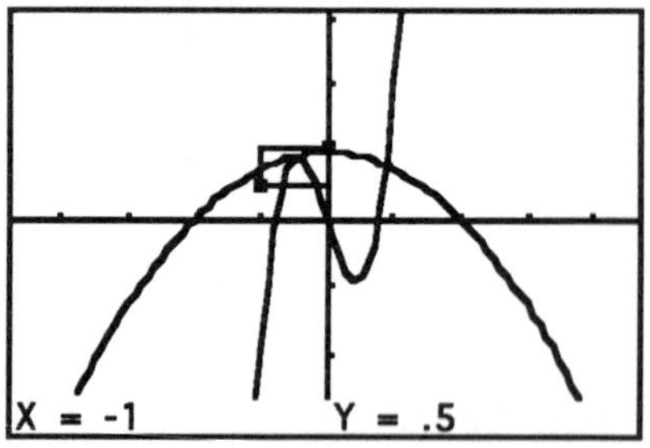

Now move that box to the point indicated in the figure above right. Notice as you move the small box a larger box forms that is enclosing the point of interest. Once you have enclosed the point with the box press [ENTER] and that small box is blown up the to size of the screen, thus zooming in on the point. The following figure reveals to us that the graphs do not intersect in the second quadrant.

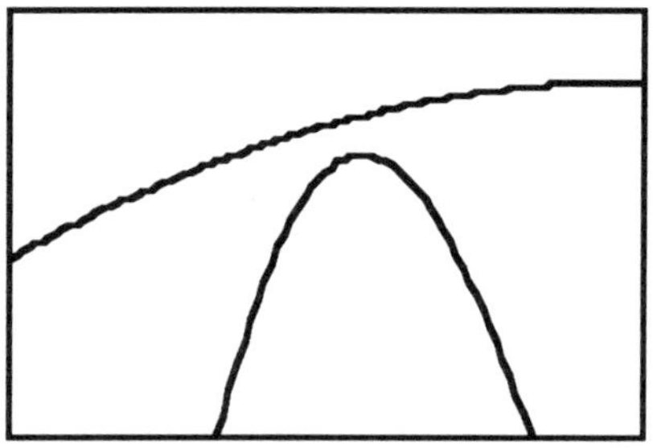

The original graph shown above indicates a point of intersection in the first quadrant. So without resetting the range press **TRACE** and hold down **▶**, scrolling the entire graph to the left. As in the previous example, we eventually will come to the other intersection point. The estimate we get this way is **X = .88297872, Y = .80508714**. These values will be slightly different on the TI-81. The graph is shown below. Now go to the **ZOOM** menu and select and enter **Set Factors**. Select **MEMORY** and then **SetFactors...** on the

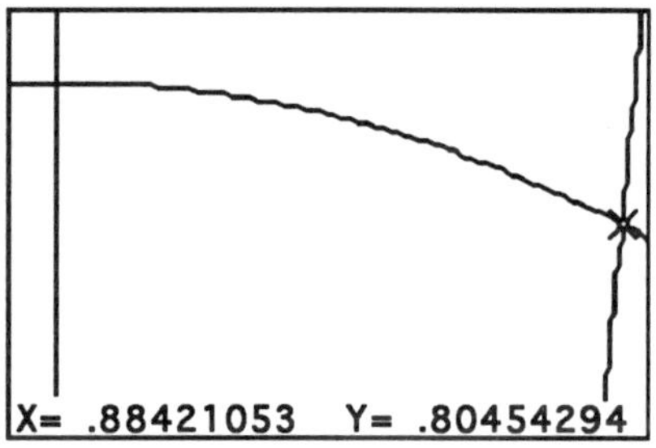

TI-82. Enter the value of 10 for each factor. Return to the **ZOOM** menu, select and enter **Zoom In**. Press **ENTER** and we have zoomed in by a factor of 10. Using the trace and zoom in procedure 4 more times gives the estimate **X = .88428404, Y = .80451043**. Slighly different values will be seen on the TI-81. The differences **Xmax – Xmin** and **Ymax – Ymin** are a measure of the accuracy. Checking the range we see that this value is accurate to about 4 decimal places.

TI-81 Graphing Review

Function	Keystrokes	Explanation
Range TI-82:	**Range** **WINDOW**	Used to the set the screen coordinates **Xmin**, **Xmax**, **Xscl**, **Ymin**, **Ymax**, **Yscl**, and **Xres**. There are two built in defaults found in the **ZOOM** menu; these are **Standard**: [-10, 10] x [-10, 10], **Xscl** = **Yscl** = **Xres** = 1. Another useful setting is [-4.8, 4.7] x [-3.2, 3.1], **Xscl** = **Yscl** = **Xres** = 1. This setting is [-4.7, 4.7] x [-3.1, 3.1], **Xscl** = **Yscl** = 1 on the TI-82 and is gotten by selectin **ZDecimal** from the **ZOOM** menu.

Clear Graphics Screen	**[2nd] [DRAW] [ENTER] [ENTER]**	This command is found in the **DRAW** menu. Clears the graphics screen and readies for another graph.
Y variables list	**[Y=]**	Stores the expressions that are graphed. Must be functions of **X** when in the **Function** mode.
Graph	**[GRAPH]**	Draws the graph(s) of the function(s) stored in the Y variables list. Returns the graph to the screen after drawing. If one of the keys [◀], [▶], [▲] or [▼] is pressed after the graph is drawn or put back on the screen, a set of cross hairs is activated with its coordinates given at the bottom of the screen. This is useful when zooming in or out. These x and y values are stored in memory locations **X** and **Y**.
Trace	**[TRACE]**	Sets a flashing box on the curve, giving the coordinates of the point at the bottom of the screen. The box can be moved to the left or right with the keys [◀] or [▶] and moved from one curve to the other by using the keys [▲] or [▼]. These x and y values are stored in memory locations **X** and **Y**.
Box	**[ZOOM] [ENTER]**	First entry in the **ZOOM** menu. Allows you to surround a part of a graph with a box and the blows that box up to the screen size. Using the keys [◀], [▶], [▲] or [▼] move one corner to the desired point and then press **[ENTER]**. This sets one corner. Then move to the opposite corner and press **[ENTER]** to zoom in on the surrounded portion of the screen.

Set Factors	ZOOM ▼ ▼ ▼ ENTER	Fourth entry in the **ZOOM** menu. Sets the zoom factors. The numbers entered must be greater than 1.
TI-82:	ZOOM **MEMORY SetFactors...**	
Zoom In	ZOOM ▼ ENTER	Second entry in the **ZOOM** menu. Puts a flashing dot at the screen center that you move to the desired place. Pressing ENTER will then zoom in on that point by the factors set in **Set Factors**.
Zoom Out	ZOOM ▼ ▼ ENTER	Third entry in the **ZOOM** menu. Puts a flashing dot at the screen center that you move to the desired place. Pressing ENTER will then zoom out on that point by the factors set in **Set Factors**.

1.5 Exercises

Draw the graph of the following functions using the given range settings. Using the trace and zoom find approximate values for the x-intercepts.

1. $y = 2x + 3$, $[-3, 3] \times [-4, 9]$.

2. $y = 4x^2 - 7x - 15$, $[-2, 4] \times [-20, 5]$.

3. $y = x^2 - 2|x|$, $[-3, 3] \times [-2, 5]$.

(The absolute value function on the CASIO is the shifted x^y key, on the fx-7700G it is in the **MATH** menu under **NUM**, and on the TI-81 it is the second function of the x^{-1} key.)

4. $y = x^3 - 2x + 1$, Default range setting.

There are two ways solve equations of the kind $f(x) = g(x)$ graphically. One way is to draw the graphs of $y = f(x)$ and $y = g(x)$ and find the points where the two graphs intersect. The other way is to draw the graph of $y = f(x) - g(x)$ and find the x-intercepts of the graph. For problems 5-7, solve each of the equations below graphically using both methods with zoom and trace techniques.

5. $|2x - 3| = 5.$

6. $3x^2 - 6x = \frac{1}{2}x + 3.$

7. $x = \frac{1}{2}\left(x + \frac{5}{x}\right).$

8. In exercise 2 of section 1.4 you found an approximate value of the radius of a cylinder of height 5 that has a volume of 1000. Construct an equation that will allow you to find the value of the radius graphically.

Investigating Order of Operations

1. Using your calculator simplify $\frac{2^3-5}{5-2}$. If you entered 2^3 - 5 ÷ 5 - 2, then the calculator answer is 5. However, the correct answer is 1. Investigate the correct way to enter the expression into the calculator by doing the following two problems as well as others you design.

A. $\dfrac{5-3^2}{5+(-6)}$ B. $\dfrac{8^2+(3)(4)}{5-3^2}$

Explain how to enter rational expressions into the calculator to get the correct answers.

2. Simplify 5 - 6 ÷ (3)(2) using order of operations. If you entered 5 - 6 ÷ (3)(2), then the calculator answer is 4. However, the correct answer is 1. Since (3)(2) means 3*2, if you enter the problem into the calculator as 5 - 6 ÷ 3 * 2, the calculator answer is 1.

Writing 3 times 2 as (3)(2) is called implicit multiplication, whereas 3*2 is called explicit multiplication. The calculator treats the two types of multiplication differently. Investigate where implicit multiplication belongs in the order of operations by doing the following four problems as well as others you design.

A. (3)(4) ÷ (2)(3) B. 3 * 4 ÷ (2)(3)

C. (3)(4) ÷ 2 * 3 D. 3 * 4 ÷ 2 * 3

Write a new order of operations including implicit multiplication.

Chapter 2

Linear Equations and Inequalities

In this chapter we will investigate ways that the graphic calculator can aid in the solving of linear equations and inequalities. In the last chapter we learned how to graph functions and using tracing and zooming, found the x-intercepts of the functions. Building on these techniques we will learn about quadratic equations and inequalities.

2.1 A Coordinate System (McKeague: Section 5.1)

Each of the graphics calculators has the facility to plot points and draw lines joining these points on a screen coordinate system. In this section we will discover how this is done and use this to draw geometric figures and find distances. We begin by giving range settings where each pixel represents a distance of one unit. These settings are:

CASIO: [-47, 47] x [-31, 31], **Xscl** = **Yscl** = 10.
TI-81: [-48, 47] x [-32, 31], **Xscl** = **Yscl** = 10.
TI-82: [-47, 47] x [-31, 31], **Xscl** = **Yscl** = 10.

Enter these range settings and then view the graphics screen. On the TI-81/82, first unselect all of the expressions in the Y variables list. To place a dot or crosshairs on the screen simply press:

CASIO: [SHIFT] [Plot] [EXE]
fx-7700G: [SHIFT] [F3] (**Plot**) [EXE].
TI-81/82: Press any arrow key with the graphics screen active.

Move the dot or crosshairs around the screen with the arrow keys, checking both the x- and y-coordinates. What do you discover? Let's look at point plotting and line drawing in detail for each calculator.

CASIO

The command sequence [SHIFT] [Plot] [EXE] (fx-7700G: [SHIFT] [F3] (**Plot**) [EXE]) places a movable dot or crosshairs at the screen center. The dot will remain in the position it was moved to until the graphics screen is cleared or the range is reset. A dot can be placed anywhere on the screen by the sequence: [SHIFT] [Plot] (number) [SHIFT] [,] (number) [EXE] (fx-7700G: [SHIFT] [F3] (**Plot**) (number) [SHIFT] [,] (number) [EXE]). As before the point can be moved around the screen with the arrow keys. For example, [SHIFT] [Plot] [(-)] [5] [SHIFT] [,] [6] [EXE] (fx-7700G: [SHIFT] [F3] [SHIFT] [(-)] [5] [SHIFT] [,] [6] [EXE]) puts a dot at the coordinate (-5, 6).

To draw lines simply set two dots on the screen and then enter **SHIFT** **Line** **EXE** (fx-7700G: **SHIFT** **F4** (**Line**) **EXE**). This will draw a line between the two points.

• <u>Example 1:</u> Draw a line between the points (-5, -10), and (8, 17).

Enter **SHIFT** **Plot** **EXE** (fx-7700G: **SHIFT** **F3** (**Plot**) **EXE**) and move the dot to (-5, -10). Press **EXE** and again another dot will be placed at the screen center. Move this one to (8, 17) and enter **SHIFT** **Line** **EXE** (fx-7700G: **SHIFT** (**Line**) **F4** **EXE**). The screen will look something like the following.

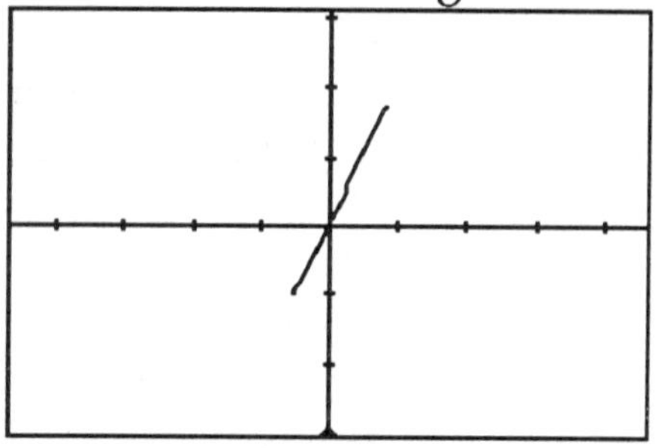

Clear the screen now. To draw geometric figures enter the following on the text screen.

Keystrokes	Explanation
Plot ◢	Sets a point at the the screen center and halts so the point can be moved.
Line	Draws a line between the points plotted. If no previous point is plotted, no line is drawn.

Press **EXE** and a dot appears at the screen center. Move it to any place on the screen and press **EXE** twice. Another dot will appear. Move this one to some point and press **EXE**. A line will then be drawn between the two points. Pressing **EXE** again puts another dot at the center. Continuing in this way it is possible to draw geometric figures. For example, in succession, select the points (0, 20), (-10, -20), (20, 5), (-20, 5), (10, -20), and (0, 20). What figure is drawn?

TI-81/82

In the **DRAW** menu of the TI-81 there are several point plotting commands. For our purposes, the third command will do nicely. On the TI-82 we will use the first command in **DRAW** menu under the **POINTS** title. First clear the graphics screen and make the graphics screen active. Immediately, go to the **DRAW** menu or **DRAW POINTS** menu on the TI-82, select and enter the **PT-On(** command. The graphics screen will return with a crosshairs. Move the crosshairs to some point and press [ENTER]. Moving the crosshairs again reveals a dot. In this way points can be plotted on the screen. Drawing lines can be accomplished in similar fashion using the **Line(** command.

Caution: **If the graphics screen is not activated before the commands PT-On(or Line(are entered a syntax error will result.**

The **PT-On(** command can also be used with a point specified. For example, make the text screen active and select and enter **PT-On(** followed with the keystrokes [(-)] [5] [ALPHA] [,] [6] [)] [ENTER]. The TI-82 has a [,] key so the keystrokes become [(-)] [5] [,] [6] [)] [ENTER]. This will plot the point (-5, 6). A line can be drawn between two points by following the command **Line(** with the coordinates of the two points.

- **Example 1:** Draw a line between the points (-5, -10) and (8, 17).

This can be done on the TI-81/82 in one of two ways.

Method 1:

Activate the graphics screen and then from the **DRAW** menu select and enter the **Line(** command. Move the crosshairs to (-5, -10) and press [ENTER]. A small square appears at the point (-5, -10). Now move the square to the point (8, 17). As this is done the line is drawn.

Method 2:

Activate the text screen and select and enter the **Line(** command followed with the keystrokes [(-)] [5] [ALPHA] [,] [(-)] [1] [0] [ALPHA] [,] [8] [ALPHA] [,] [1] [7] [)] [ENTER]. On the TI-82 the keystrokes are [(-)] [5] [,] [(-)] [1] [0] [,] [8] [,] [1] [7] [)] [ENTER]

Either method produces the picture

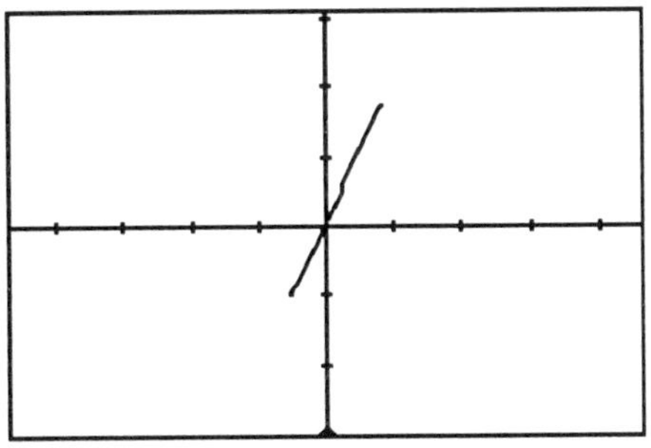

Using method 1, draw lines between the points (0, 20), (-10, -20), (20, 5), (-20, 5),

(10, -20), and (0, 20) in succession. What figure is drawn?

Techniques similar to the above can be used in programs to plot points that are input from the keyboard or generated by a formula. Before we look at this we will first give a synopsis of the necessary program commands used for loops.

CASIO

Key	Keystroke	Position and Meaning
Lbl	[SHIFT] [Lbl] (number)	Found above the [⇐] key. Labels a line or section of commands and is followed by a number in the range 0-9. Used in conjunction with Goto.
fx-7700G	[SHIFT] [PRGM] [F1] (JMP) [F3] (Lbl)	
Goto	[SHIFT] [Goto] (number)	Found above the [Prog] key. Causes the program to jump to the line labeled with a number. Followed by a number in the range 0-9. Used in conjunction with Lbl.
fx-7700G	[SHIFT] [PRGM] [F1] (JMP) [F2] (Gto)	

TI-81/82

Key or Command	Keystroke	Position and Meaning
Lbl	[PRGM] [ENTER] (number)	Found in the **CTL** menu when on line 2 or greater of the edit screen for a program. Labels a line or section of commands and is followed by a number in the range 0-9. Used in conjunction with Goto.
TI-82	9th item in **CTL** menu	
Goto	[PRGM] [▼] [ENTER] (number)	Found in the **CTL** menu when on line 2 or greater of the edit screen for a program. Causes the program to jump to the line labeled with a number. Followed by a number in the range 0-9. Used in conjunction with Lbl.
TI-82	10th item in **CTL** menu	

• **Example 3:** Plot the points given in the following table.

x	-4.2	-3.1	-1.5	0.3	2.0	2.7	4.1
y	2.3	1.9	1.3	0.94	0.67	0.58	0.44

It is easily seen that if the range is set to [-4.5, 4.5] x [0, 2.5] all of the points will appear on the screen.

CASIO

The following program takes input from the keyboard and plots the points on the screen.

```
Lbl 1:"X"?→X ↵
"Y"?→Y ↵
Plot X,Y ↵
Goto 1 ↵
```

Execute the program and enter each pair **X, Y** in succession. When all of the values are entered press [MODE] [1] to halt the program. Press [AC] and access the graphics screen to see

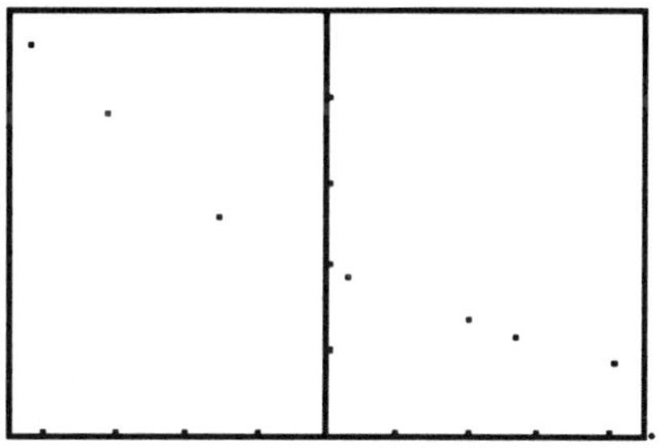

TI-81/82

The following program takes input from the keyboard and plots the points on the screen.

```
:Lbl 1
:Disp "X"
:Input X
:Disp "Y"
:Input Y
:PT-On(X,Y)
:Goto 1
```

Execute the program and enter each **X, Y** pair in succession. When all of the values are entered press **ON** and select **Quit**. Activate the graphics screen to reveal the graph shown below.

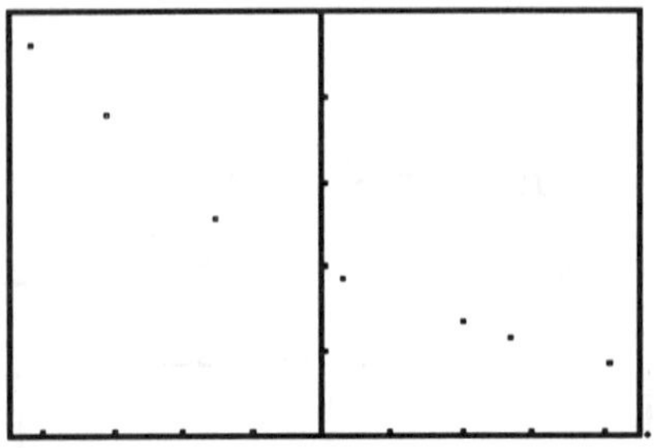

By adding a few changes to the above programs points can be plotted that are generated with an algebraic formula. Simply replace the lines where **Y** is entered with the formula. For example, entering the expression **2X-1→Y** will plot points given by the expression $y = 2x - 1$.

Before looking at how this might be done we give a program that sets the screen with a nice coordinate system. One that has the x-axis and the y-axis scaled in ratio of 1:1. The following programs are useful utilities and should be kept in a convenient program storage location.

Program 2.1.1C: Range Setting Utility

Keystrokes	Explanation
"X"?→X:"Y"?→Y ↵	Prompts for the input of the center point.
"S"?→S ↵	Prompts for the input of the scale. This will be the increment between each pixel.
Range X–47S,X+47S,10S, Y–31S,Y+31S,10S ↵	Sets the range.

TI-81/82

The changes needed for the TI-82 are indicated in parentheses.

Program 2.1.1T: RANGE Range Setting Utility

Commands	Explanation
:Disp "X"	Prompts for the input of the center point.
:Input X	
:Disp "Y"	
:Input Y	
:Disp "S"	Prompts for the input of the scale. This will be the increment between each pixel.
:Input S	
:X–48S→Xmin (X-47S→Xmin)	Sets the range.
:X+47X→Xmax	
:10S→Xscl	
:Y–32S→Ymin (X-31S→Ymin)	
:Y+31S→Ymax	
:10S→Yscl	

The only requirement of these programs is that **S** be positive.

- <u>Example 4:</u> Plot the points generated by the formula y = 2x – 3 using the following numbers for x, -2, -1, 0, 1, 2, 4. (McKeague: Example 3, 5.1)

Some prelinimary arithmetic tells that when x = -2, y = 2(-2)-3 = -7 and when x = 4, y = 2(4)-3 = 3. So we will set the screen range with program 2.1.1C or 2.1.1T using as input **X = Y = 0** and **S = .25**.

CASIO

Enter the following program. An additional command is put in to display the value of y before the pair (x, y) is plotted.

```
Lbl 1:"X"?→X ↵
2X–3→Y ↵
"Y":Y ◢
Plot X,Y ↵
Goto 1 ↵
```

TI-81/82

Enter the following program. An additional command is put in to display the value of y before the pair (x, y) is plotted.

```
:Lbl 1
:Disp "X"
:Input X
```

```
:2X–3→Y
:Disp "Y"
:Disp Y
:PT-On(X,Y)
:Goto 1
```

Executing the program an entering in succession the values of x (-2, -1, 0, 1, 2, 4) produces the graph shown below. The corresponding y values are -7, -5, -3, -1, 1, 5.

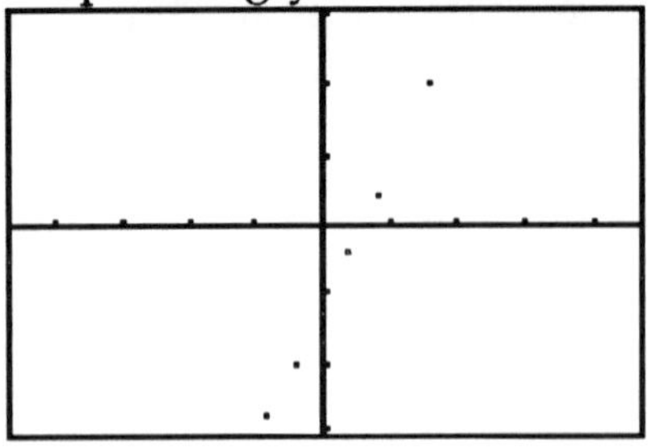

2.1 Exercises

Plot the given points and draw a line joining the points.

1. (-5, 1), (4, 8).
2. (-3, 3), (-8, -1).
3. $(\sqrt{2}, 0), (1,1)$.
4. $\left(\frac{3}{5}, \frac{4}{5}\right), \left(-\frac{4}{5}, \frac{3}{5}\right)$.

5. Plot the points given in the following table

x	0	.1	.2	.3	.4	.5	.6	.7	.8	.9	1
y	4	2.93	1.01	-.92	-2.17	-2.43	-1.78	-.61	.55	1.32	1.47

6. Refer to exercise 3, section 1.4. Create a table of the ordered pairs $(\mathbf{R}, \mathbf{V}/\mathbf{S})$ for several values of **R** and **H** where **R** = c**H**, c > 0. Plot these ordered pairs. What is the relationship between **R** and **V/S**?

Plot the points generated by the following formulas. Use the following numbers for x. -5, -4, -3, -2, -1, 0, 1, 2, 3, 4, 5. Which have the pairs (x, y) lying on a straight line?

7. $y = 1 - 2x$.
8. $y = -1$.
9. $y = 3 - \frac{1}{2}x$.
10. $y = x - \frac{1}{2}x^2 + 1$

11. The following table is an amortization schedule for a 1 year car loan of $5000 with an annual interest rate of 9.9% and monthly payments of $439.35.

Payment No.	Payment	Interest	Unpaid Balance Reduction	Unpaid Balance
0				5000
1	439.35	41.25	398.10	4601.90
2	439.35	37.97	401.38	4200.52
3	439.35	34.65	404.69	3795.83
4	439.35	31.32	408.03	3387.80
5	439.35	27.95	411.40	2976.40
6	439.35	24.56	414.79	2561.61
7	439.35	21.13	418.21	2143.40
8	439.35	17.68	421.66	1721.73
9	439.35	14.20	425.14	1296.59
10	439.35	10.70	428.65	867.94
11	439.35	7.16	432.19	435.75
12	439.35	3.59	435.75	0
Total	5272.16	272.16	5000.00	

Plot the pairs (Payment No., Interest), (Payment No., Reduction), and (Payment No., Unpaid Balance). Does it appear that these pairs lie on a straight line? Should they lie on a line?

2.2 Linear Equations

(McKeague: Sections 2.1 – 2.5)

A linear equation is one that can be put in the form ax + b = c. This equation can be visualized as the two expressions y = ax + b and y = c and what we are trying to do is find all of the values of x where that produces the same y values. In the previous section we noticed that the set of points that satisfy the equations y = ax + b and y = c when plotted on a coordinate system produce straight lines. With this in mind we could solve the equation ax + b by finding where the lines y = ax + b and y = c cross.

• **Example 1:** Solve the equation $\frac{3}{4}x + 5 = -4$, graphically and algebraically.

(McKeague: 2.1, Example 4)

CASIO

First set the range to be [-23.5, 23.5] x [-15.5, 15.5], **Xscl** = **Yscl** = 5. This can be obtained using program 2.1.1C and entering **X = Y = 0** and **S** = 0.5. Enter **Graph Y=-4:Graph Y=3X÷4+5** and press [EXE]. After the graphs are drawn press [SHIFT] [Trace] ([F1] on the fx-7700G) and move the flashing dot or cross hairs to the point where the lines meet. You should find that **X = -12** and **Y = -4**.

Checking we find

$$\frac{3}{4}(-12) + 5 = 3(-3) + 5 = -9 + 5 = -4.$$

This verifies that the solution is x = -12. The graph of the situation is shown below.

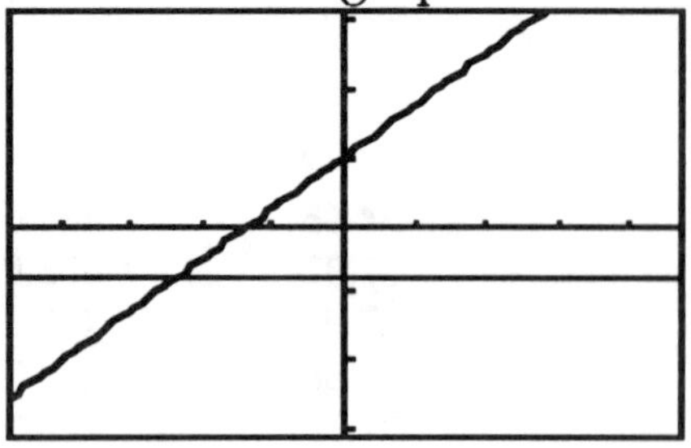

TI-81/82

First set the range to be TI-81: [-24, 23.5] x [-16 15.5], **Xscl** = **Yscl** = 5 or TI-82: [-23.5, 23.5] x [-15.5, 15.5], **Xscl** = **Yscl** = 5. In the **Y-VARS** menu enter **3X/4+5** for **Y1** and **-4** for **Y2**, then press **GRAPH**. After the graphs are drawn press **TRACE** and move the cross hairs the point where the lines meet. You should find that **X = -12** and **Y = -4**. Checking we find

$$\frac{3}{4}(-12) + 5 = 3(-3) + 5 = -9 + 5 = -4.$$

This verifies that the solution is x = -12. The graph of the situation is shown above. Algebraically, this equation is solved via the following steps.

$$\frac{3}{4}x + 5 = -4$$

$$\frac{3}{4}x + 5 - 5 + -4 - 5$$

$$\frac{3}{4}x = -9$$

$$\frac{4}{3}\left(\frac{3}{4}x\right) = \frac{4}{3}(-9)$$

$$x = -12.$$

As you can see, a graph can be used to solve these types of equations. However, there are times when exact solutions may be difficult to find from a graph. In those cases, the graph helps to find estimates of the solution and so that is ultimately no substitute for finding solutions algebraically.

• <u>**Example 2:**</u> Solve 10x – 5 = 3x – 2. Estimate the solution with a graph and find the exact solution algebraically.

In this case we are searching for the point where the graphs of y = 10x – 5 and y = 3x – 2 meet.

CASIO

Set the range to be the default setting [-4.7, 4.7] x [-3.1, 3.1] and enter **Graph Y=10X–5:Graph Y=3X–2** and press **EXE**. You should see the graph at the top of the next page.

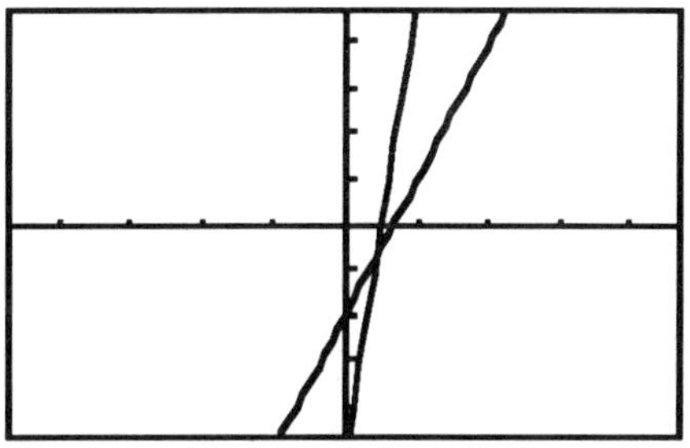

Pressing **SHIFT** **Trace** (**F1** on the fx-7700G) and moving the flashing dot or cross hairs to the point where the graphs meet we find **X = 0.5** and **Y = -0.5**. Is the value of x the solution? Let's check.

$$10(0.5) - 5 = 5 - 5 = 0$$
$$3(0.5) - 2 = 1.5 - 2 = -0.5$$

We see from this computation that x = 0.5 is not a solution. In order to get a better estimate we can zoom in on this point. Press **SHIFT** **X** (**F2** **F3** on the fx-7700G). Make certain that the zoom factors on the fx-7700G both equal to 2. (See the previous chapter to recall how this is done. Now that we have zoomed in, the graph now looks like the figure at the top of the next page.

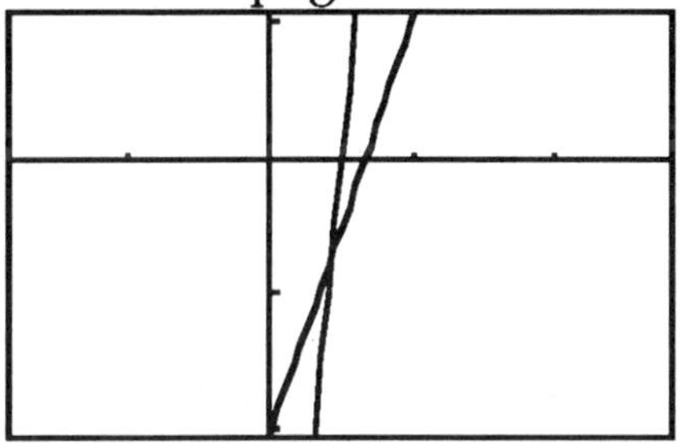

Using the trace we find the estimate for the solution to be **X = 0.45** and **Y = -0.65**. Let's check this estimate again.

$$10(0.45) - 5 = 4.5 - 5 = -0.5$$
$$3(0.45) - 2 = 1.35 - 2 = -0.65$$

This shows that our estimate is better, because the y values we get are closer than before. We could keep zooming in until we more accurate estimates, but at least we now know that the solution is near 0.45.

TI-81/82

Set the range to the default setting, TI-81: [-4.8, 4.7] x [-3.2, 3.1], TI-82: [-4.7, 4.7] x [-3.7, 3.1]. Enter in **Y1** the expression **3X–2**, in **Y2** enter **10X–5**, and press **GRAPH**. You should see the graphs shown at the top of the next page.

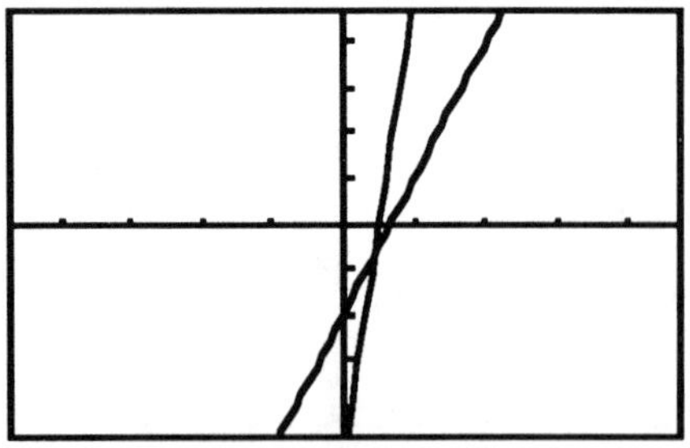

After the graphs are drawn press **TRACE** and move the cross hairs to the point where the graphs meet we find **X = .5** and **Y = -.5**. Is the value of x the solution? Let's check.

$$10(0.5) - 5 = 5 - 5 = 0$$
$$3(0.5) - 2 = 1.5 - 2 = -0.5$$

We see from this computation that x = 0.5 is not a solution. In order to get a better estimate we can zoom in on this point. Set the zoom factors both equal to 2. (See the previous chapter to recall how this is done.) Go to the **ZOOM** menu and select **Zoom In**. Now that we have zoomed in, the graph now looks like the following figure.

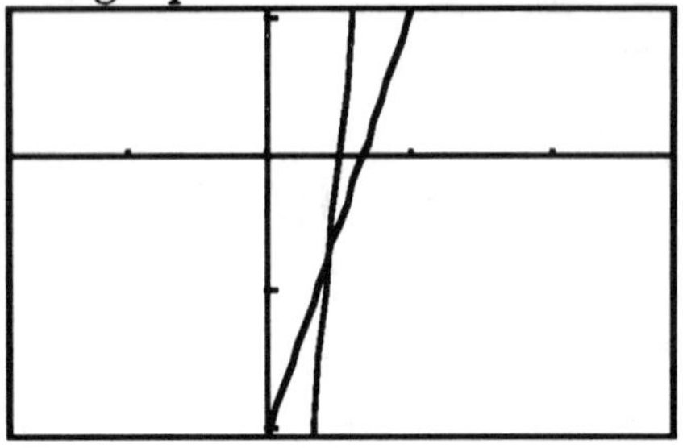

Using the trace we find the estimate for the solution to be TI-81: **X = 0.425** and **Y = -0.725** TI-82: **X = 0.45** and **Y = -0.65**. Let's check these estimates again.

TI-81	TI-82
$10(0.425) - 5 = 4.25 - 5 = -0.75$	$10(0.45) - 5 = 4.5 - 5 = -0.5$
$3(0.425) - 2 = 1.275 - 2 = -0.725$	$3(0.45) - 2 = 1.35 - 2 = -0.65$

This shows that our estimate is better, because the y values we get are closer than before. We could keep zooming in until we more accurate estimates, but at least we now know that the solution is near 0.45.

Algebraically we find

$$10x - 5 = 3x - 2$$
$$10x - 5 + 5 = 3x - 2 + 5$$
$$10x = 3x + 3$$
$$10x - 3x = 3x - 3x + 3$$
$$\frac{1}{7}(7x) = \frac{1}{7}(3)$$
$$x = \frac{3}{7}.$$

This shows that the solution is $x = \frac{3}{7}$ which is approximately 0.42857.

2.2 Exercises

Solve each of the following equations graphically and algebraically. If a solution cannot be found exactly with a graph, find an estimate that is accurate to 2 decimal places.

1. $8 - 2x = 6$.
2. $7 = 4 + \frac{1}{3}x$.
3. $3x + 4 = -2x - 6$.
4. $-x + 7 = 5 - 11x$.
5. $2(3x - 3) + 5 = 9x + 2$.
6. $0.75x + 1.05 = -0.4x + 0.13$.

2.3 Inequalities

(McKeague: Sections 2.6, 2.7, 2.8)

Linear inequalities are expressions of the type

$$ax + b < c \text{ or } |ax + b| \geq c.$$

The goal here is to develop a graphical meaning to these expressions and determine their solution sets.

The solution of the inequality $ax + b < c$ can be visualize as finding where the graph of $y = ax + b$ is below the graph of $y = c$. Of course if the symbol '≥' is used we would find where the graph of $y = ax + b$ is above or equal to the graph of $y = c$. Let's illustrate these ideas with some examples.

• <u>Example 1:</u> Solve the inequality $3x - 5 \leq 7$. (McKeague: Example 2, Section 2.6)

CASIO

Set the range using the program 2.1.1C, with **X = Y = 0** and **S = .5**. This will give the range setting [-23.5, 23.5] x [-15.5, 15.5], **Xscl = Yscl = 5**. Once the range is set enter **Graph Y = 7:Graph Y = 3X–5**, and press **EXE**. This produces the graph

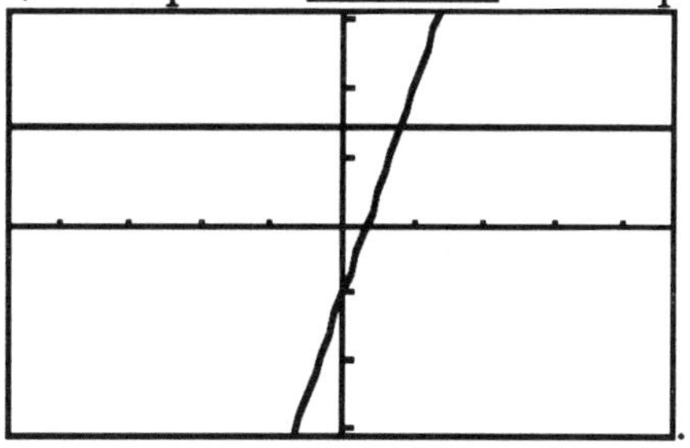

TI-81/82

Set the range using the program 2.1.1T, with **X = Y = 0** and **S = .5**. This will give the range setting TI-81: [-24, 23.5] x [-16, 15.5], TI-82: [-23.5, 23.5] x [-15.5, 15.5], **Xscl = Yscl = 5**. In the **Y=** menu enter **3X–5** for **Y1**, **7** for **Y2** and press **GRAPH**. The graph produced should look as the one depicted above.

In solving this inequality we want to find all of the values of x where the graph of $y = 3x-5$ is below or is equal to the graph $y = 7$. Looking at the graph we see this solution set to be all of the x values left of the point where the lines meet. Activating the trace and moving to this point we find **X = 4** and **Y = 7**. So it appears that the solution is given by the set

$$\{x \mid x \le 4\} = (-\infty, 4].$$

Solving this algebraically produces

$$3x - 5 + 5 \le 7 + 5$$
$$3x \le 12$$
$$\tfrac{1}{3}(3x) \le \tfrac{1}{3}(12)$$
$$x \le 4.$$

• **Example 2:** Solve the inequality $-2 < 4 - 5x < 3$.

In this example we are searching for the x values where the graph of $y = 4 - 5x$ lies above $y = -2$ and below the graph of $y = 3$.

CASIO

Set the screen range to the default setting. Enter the following

Graph Y=-2
:Graph Y=3
:Graph Y=4–5X

and then pressing **EXE**, produces the graph at the top of the next page. It is seen that the solution to the inequality is the set of x values that lies between the points where the

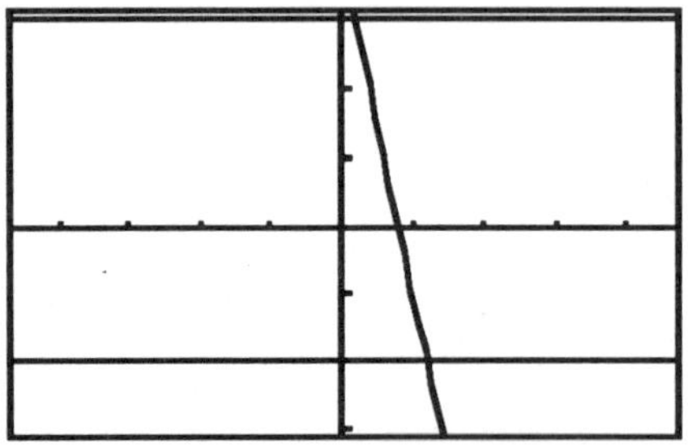

graphs meet. Activating the trace and moving to these points we find the solution to be the set $\{x \mid 0.2 < x < 1.2\} = (0.2, 1.2)$.

TI-81/82

Set the screen range to be the default and in the **Y=** menu enter **4–5X** for **Y1**, **3** for **Y2**, and **-2** for **Y3**. Pressing **GRAPH** produces the graph dipicted above. The graphs shows that the solution is the set of x values that lie between the point where the graphs meet. Activating the trace and moving to these points we find the solution to be the set

$$\{x \mid 0.2 < x < 1.2\} = (0.2, 1.2).$$

Solving this inequality algebraically produces

$$-2 < 4 - 5x < 3$$
$$-2 - 4 < 4 - 4 - 5x < 3 - 4$$
$$-6 < -5x < -1$$

$$-\frac{1}{5}(-6) > -\frac{1}{5}(5x) > -\frac{1}{5}(-1)$$
$$\frac{6}{5} > x > \frac{1}{5} \text{ or } \frac{1}{5} < x < \frac{6}{5}.$$

The absolute value of a number is the distance that number is from zero. so since $|x|$ is distance, it must be the case that $|x| \geq 0$. This tells us that the graph of $y = |x|$ lies entirely above the x axis except at the point $x = 0$. So in dealing with inequalities involving absolute value the graphs we view will be above the x-axis. Let's make this more precise with the following example.

• <u>Example 3:</u> Solve the inequality $|2x - 5| \geq 2$.

This is equivalent to

$$2x - 5 \leq -2 \text{ or } 2x - 5 \geq 2$$
$$x \leq \frac{3}{2} \text{ or } x \geq \frac{7}{2}.$$

CASIO

Set the range to the default and draw the graphs $y = |2x - 5|$ and $y = 2$ on the same screen. This is accomplished by entering

Graph Y=2

:Graph Y=Abs(2x–5)

and then pressing **EXE**. This produces the following graph shown here.

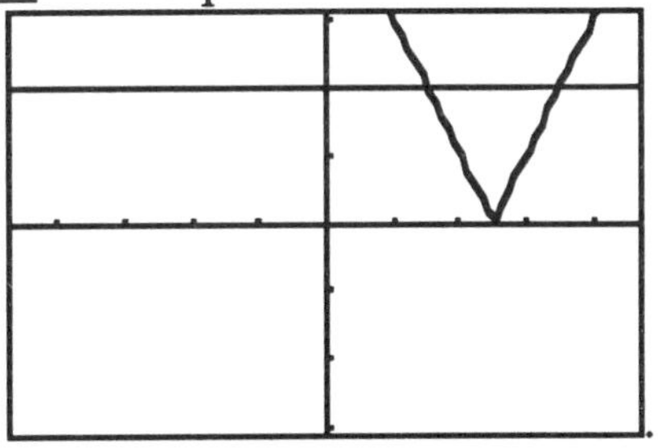

We want to find all the values of x where the graph of $y = |2x - 5|$ is above or meets the graph of $y = 2$. We see from the graph that these values are to the left of the first point where the graphs meet and to the right of the second such point. Using the trace we find the intersection points to be approximately 1.5 and 3.5. At these points the y coordinate is 2. From the algebra given above we know that these are in fact the exact values.

On the CASIO it should be advised to plot the graphs in different orders since the trace is only active on the last graph drawn. As you trace both x and y coordinates should also be viewed. Both curves can be traced with the fx-7700G.

TI-81/82

Set the range to the default. Enter **abs(2X-5)** into $\mathbf{Y_1}$ and **2** into $\mathbf{Y_2}$ in the Y variables list and draw the graphs. This will produce the graph at the top of the next page

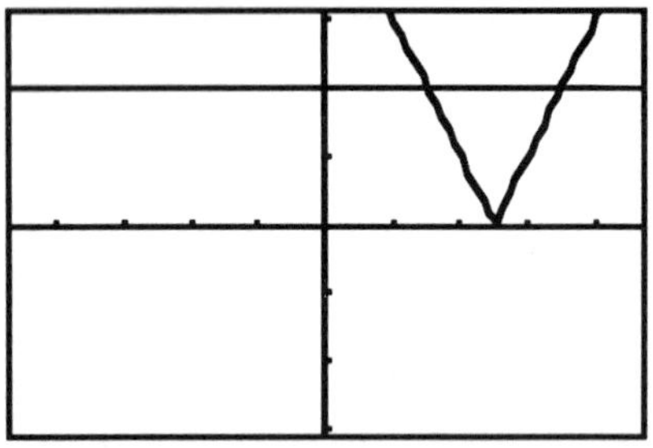

Using the trace and toggling between the two graphs we find that $x \le 1.5$ or $x \ge 3.5$.

Caution: **Because of the precision of the calculators (or the lack thereof) it is still necessary to solve the inequalities algebraically. All the graphs provide is an estimate of the solution sets.**

The Casio fx-7700G has a feature that is useful for visualizing the solution to inequalities that shades in areas that lie above, below or between curves. This is done by entering the **INEQ** graph type (SHIFT MODE ÷). When Graph is pressed a menu with **Y>**, **Y<**, **Y≥**, and **Y≤** is displayed.

The TI-81/82 also has a shading feature. In the **DRAW** menu screen there is the command **Shade(**. This command has the format **Shade(**l(x), u(x), r, left, right**)**, where l(x) is the expression for the lower function, u(x) is the expression for upper function, r is the resolution of the shading, left is the left endpoint or function, and right is the right endpoint or function. The values or expressions for r, left and right are optional; if they are not entered, r is given the value of 1, left will be given the value **Xmin** and right will be given the value **Xmax**.

• <u>Example 3 (continued):</u> Solve the inequality $|2x - 5| \ge 2$.

CASIO

Put the calculator in the INEQ graph mode (SHIFT MODE ÷) and set the range to the default setting. To then draw a picture of the solution set of $|2x - 5| \ge 2$ enter

Graph Y≥2:Graph Y≤Abs (2X-5).

This will produce the following picture. To understand why this produces the proper

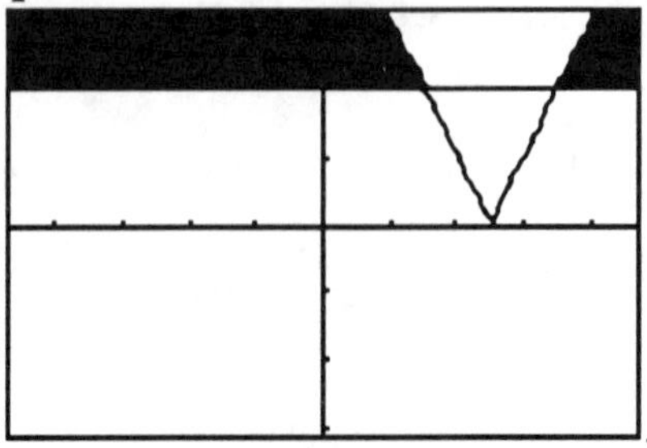

graph, recall we want to find all x values where $y = |2x - 5|$ lies above $y = 2$. So we want to find all the x values where $y \ge 2$ but $y \le |2x - 5|$.

TI-81/82
Unselect all of the functions in the **Y=** list. Set the range to the default setting and under the **DRAW** menu enterthe 7th item **Shade(**. Following this enter **abs (2X–5),2)**. The screen should have **Shade(2,abs (2X–5))**. The graph should look like that depicted above.

2.3 Exercises

Solve the following inequalities graphically and algebraically.

1. $5x + 2 > x - 3$.
2. $2x + 4 \leq 6x - 8$.
3. $\dfrac{2x - 5}{3} + 2x < \dfrac{x - 4}{2}$.
4. $|3x + 5| < 1.7$.
5. $|4 - 3x| \geq 2$.
6. $1 \leq |5 - 3x| < 5$.

Investigating Absolute Value

The following equation contains the constants a, b, c, and d. These constants are call parameters and control the number of solution(s) of the equation,

$$|x - a| + b = cx + d.$$

Investigate the possible number of solutions of this equation. Complete Table 1 to ass you in this investigation. To find the number of solutions for a given set of paramet graph each side of the equation as shown below.

$$y = |x - a| + b \qquad y = cx + d$$

The graph of the first case in the table below where a=2, b=0, c=0, and d=2, will look like TI-82 display in Figure 1.

$$y = |x - 2| \qquad y = 2$$

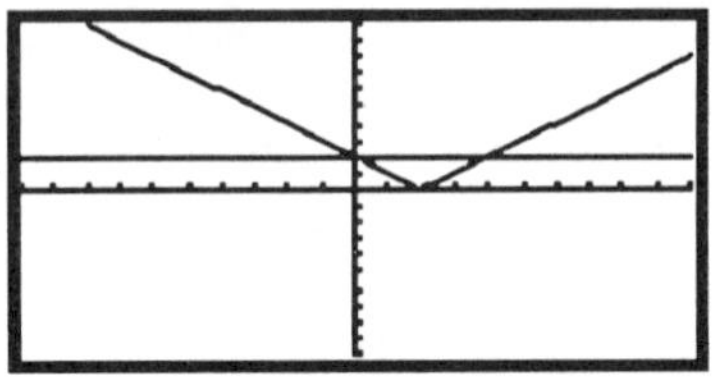

Figure 1

Note that the graphs intersect at two points. The x values of the points of intersection the solutions to the original equation.

a	b	c	d	Number of Solutions
2	0	0	2	2
2	2	0	2	
2	4	0	2	
2	2	0	4	
2	4	0	4	
2	6	0	4	
2	1	2	2	
2	4	1	2	
2	6	1	2	
2	0	-1	0	
2	-2	-1	0	
2	-4	-1	0	
2	0	-2	0	
2	0	-1/2	2	
2	4	-1/4	-2	

Table 1

Discuss the possible number of solutions. Look for relationships between the parameters and the number of solutions

Chapter 3

Polynomials and Factoring

Probably the most widely used functions are polynomials. Their use is pervasive in all areas of engineering and business. You will see them crop up in computer aided design, signal processing, data fitting, and profit analysis. In this chapter we will look at this family of functions and what the graphics calculator can do to aid us.

3.1 Polynomials

(McKeague: Sections 3.5-3.9)

A polynomial of degree n has the form

$$a_nx^n + a_{n-1}x^{n-1} + \cdots + a_1x + a_0$$

where $a_n, a_{n-1}, \cdots, a_1, a_0$ are real numbers and $a_n \neq 0$. The feature of polynomials we will investigate graphically is factoring. This we shall see is tied to values of x where the polynomial is zero or what is called the x-intercepts of the graph of

$$y = a_nx^n + a_{n-1}x^{n-1} + \cdots + a_1x + a_0.$$

The following graphs illustrate typical polynomials.

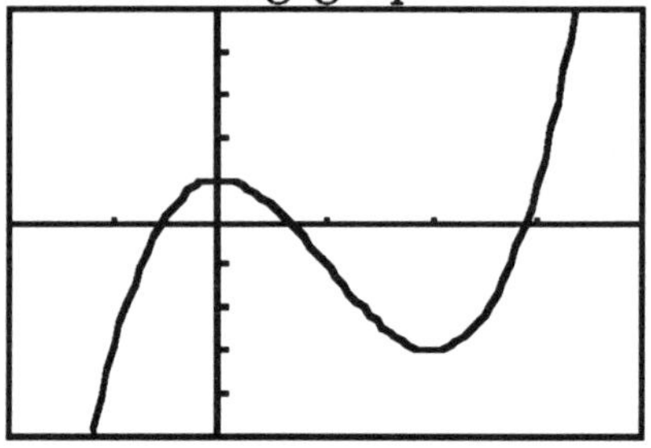

Third Degree

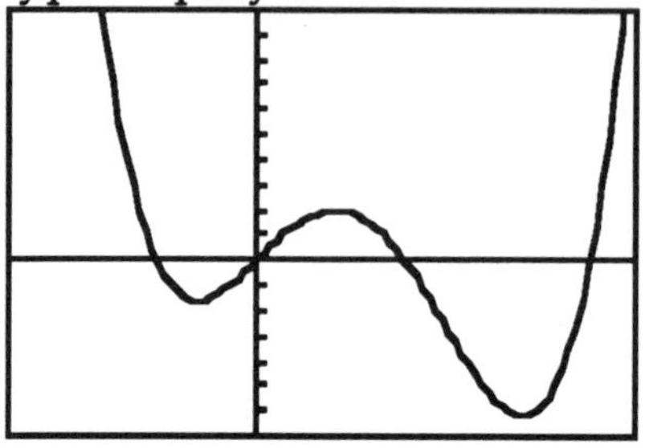

Fourth Degree

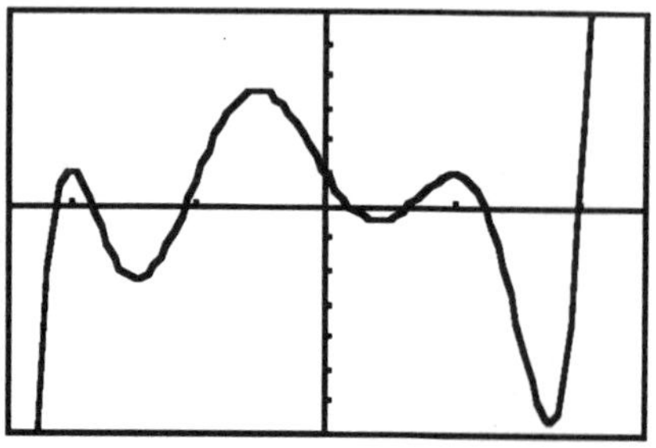

Seventh Degree

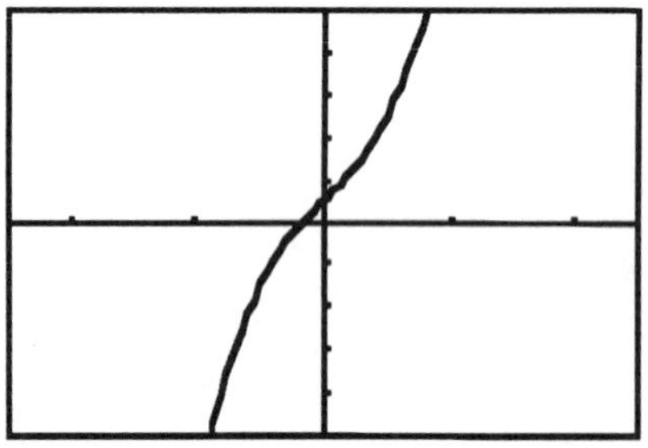

Third Degree

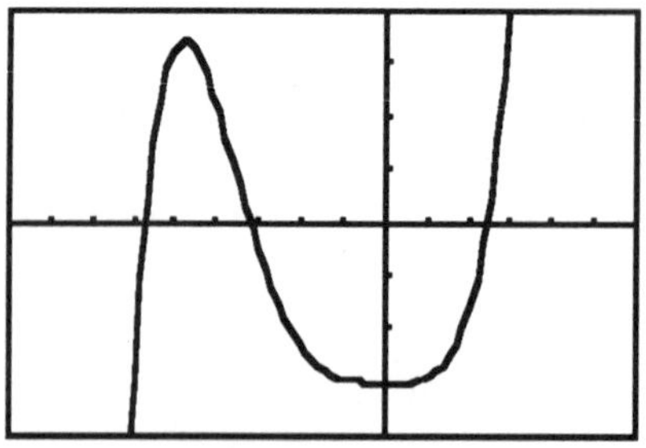

Fifth Degree

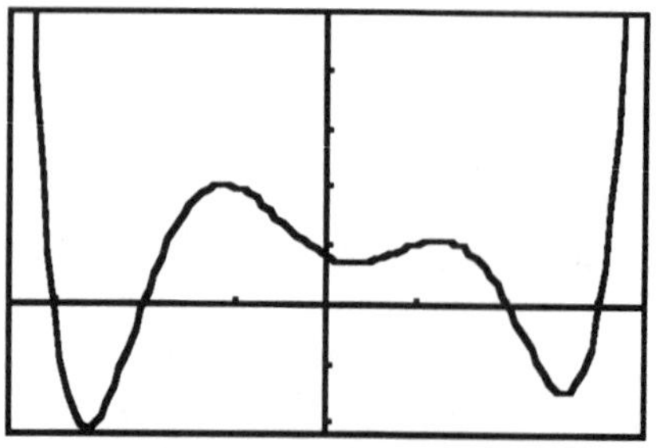

Sixth Degree

There are two results about polynomials with one variable that will help us in determining how to factor it, the Remainder Theorem and the Factor Theorem.

Remainder Theorem

If the polynomial $a_nx^n + a_{n-1}x^{n-1} + \cdots + a_1x + a_0$ is divided by $x - c$ then the remainder is $a_nc^n + a_{n-1}c^{n-1} + \cdots + a_1c + a_0$.

Factor Theorem

If, for the polynomial $a_nx^n + a_{n-1}x^{n-1} + \cdots + a_1x + a_0$,

$a_nc^n + a_{n-1}c^{n-1} + \cdots + a_1c + a_0 = 0$ then $x - c$ is a factor..

So when viewing the graph of a polynomial we will look for x values where it is zero or crosses the x-axis. These x values according to the Factor Theorem will identify the factors. Let's see how this is done with an example.

- <u>Example 1:</u> Using the graph of $y = 3x^3 - 2x^2 - 12x + 8$ find all of this polynomials factors.

CASIO

Set the range to the default setting and enter **Graph** $Y=3Xx^y3-2X^2-12X+8$ and press [EXE]. The following graph should be produced.

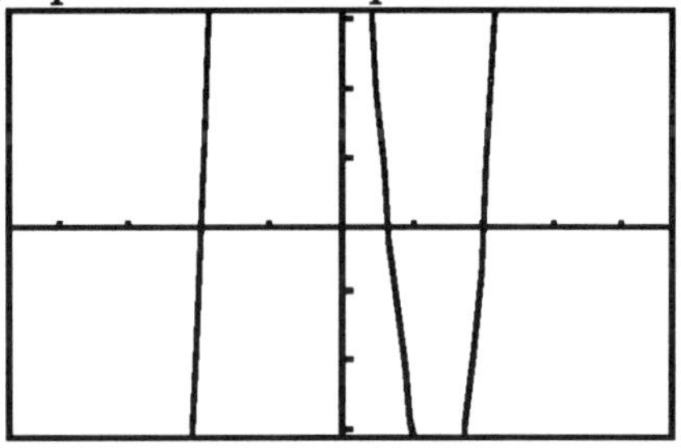

TI-81/82

Set the range to the default setting and enter $3X^3-2X^2-12X+8$ for **Y1** and press [GRAPH]. The same graph shown above should be produced.

Using the trace we see that if $x = -2$ the $y = 0$. By the Factor Theorem we know that $x + 2$ must be a factor. Also using trace we find that when $x = 2$, $y = 0$. So again the factor Theorem implies that $x - 2$ is a factor. This implis that $(x + 2)(x - 2) = (x + 2)x - (x + 2)2 = x^2 + 2x - 2x - 4 = x^2 - 4$ is factor. Trying the method of grouping gives

$$3x^3 - 2x^2 - 12x + 8 = x^2(3x - 2) - 4(3x - 2) = (3x - 2)(x^2 - 4).$$

- <u>Example 2:</u> A football is kicked straight up from a height of 4 feet above the ground with an initial upward velocity of 30 feet per second. The equaiton that gives the height of the ball at any time t is $h = -16t^2 + 30t + 4$. Factor the right side of the equation and find h when $t = 1.5$ seconds. Find also the

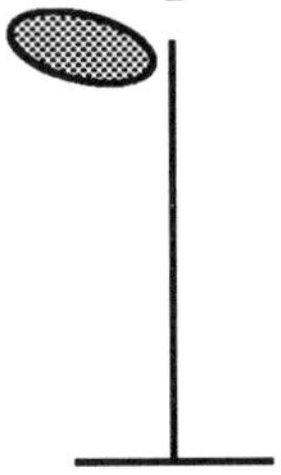

time when the football hits the ground.

CASIO

Since the CASIO graphic calculators (fx-7700G in the **REC** mode) only graph expressions involving X, we will substitute X for t. Since t represents time which is greater than or equal to zero as is the same for h, representing height we will set the range to be [0, 5] x [0, 10], **Xscl** = **Yscl** = 1. Enter then **Graph Y=-16X²+30X+4** and press **EXE**. The following graph is shown.

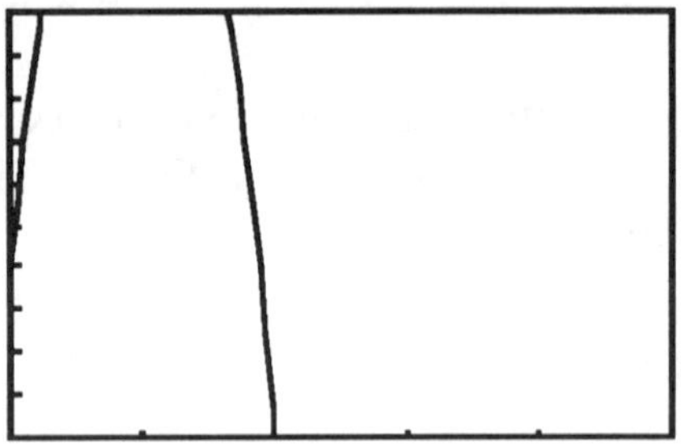

TI-81/82

The TI graphics calculators in the **Function** or **Func** mode only graph expression involving X, we will substitued **X** for t. Since t represents time which is greater than or equal to zero as is the same for h, representing height we will set the range to be [0, 5] x [0, 10], **Xscl** = **Yscl** = 1. Enter **-16X²+30X+4** for **Y1** and press **GRAPH**. The graph produced is shown above.

We see that the graph crosses the x-axis whe $x = 2$. Taking $t = 2$ we find

$$h = -16(2)^2 + 30(2) + 4 = -16(4) + 60 + 4 = -64 + 64 = 0.$$

So the Factor Theorem implies that $t - 2$ is a factor of h. Now we have that

$$-16t^2 + 30t + 4 = 2(-8t^2 + 15t + 2)$$

and from what we know about factoring trinomials, h must factor to $(t - 2)(-8t - 1)$ because $-8(1) = -8$, $(-2)(-8) + (1)(-1) = 16 - 1 = 15$, and $(-2)(-1) = 2$. Thus $h = 2(t - 2)(-8t - 1)$. When $t = 1.5$ we have $h = 2(1.5 - 2)(-8(1.5) - 1) = 2(-.5)(-13) = 13$. so the football is 13 feet above the ground after 1.5 seconds. To determine when the football hits the ground we must find when $h = 0$ and from what we noticed earlier when $t = 2$, it is clear that after 2 seconds the football has hit the ground.

By producing a graph of the polynomial, estimates of the x values the polynomial is zero appears to be quite easy, but lack of preknowledge about how to set the range to see the entire graph of the polynomial can cause some difficulties.

• <u>Example 3:</u> Locate estimates for the x-intercepts of the polynomial $p = 10x^4 - 181x^3 - 304x^2 + 181x + 294$, then use these estimates to factor p.

Set the screen to setting [-3, 3] x [-20, 20]. **Xscl** = 1, **Yscl** = 2. The graph should appear as the figure at the top of then next page.

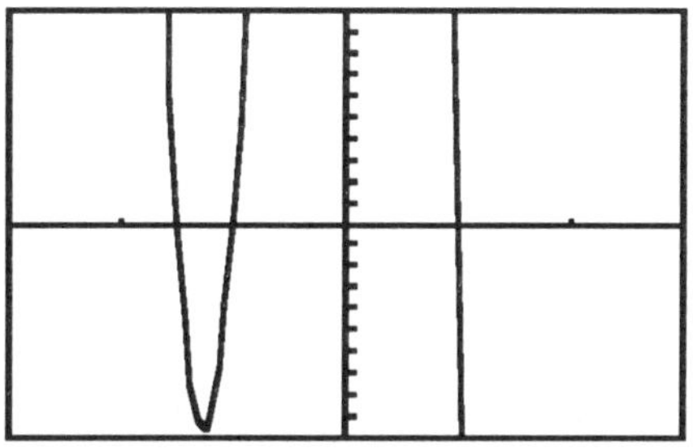

We see three of the x-intercepts but not all of them. Since this is a fourth degree polynomial and we need to factor it, the Factor Theorem tells us that there should be one more intercept. Using the trace with this graph as it is, we are not be able to estimate the intercpets to any accuracy at all, but from the way the x-axis is scaled, in this case **Xscl** = 1, we see that estimates for the roots are $-\frac{3}{2}$, -1, and 1. By evaluating the polynomial at each of these points we find that these are intercepts. The Factor Theorem says that $x + \frac{3}{2}$, $x + 1$, and $x - 1$ are factors of p. Now if $x - \frac{3}{2}$ is a factor then setting this equal to zero tells us that $x = -\frac{3}{2}$, but this is also the solution of the equation $2x - 3 = 0$. So in order to obtain the proper factorization of p, it is necessary to take the factor to be $2x - 3$ instead of $x + \frac{3}{2}$. This will aid in getting a leading coefficient of 10. Now set the range to be, [-15, 15] x [-1000, 1000], **Xscl** = 1, **Yscl** = 100 and redraw the graph which now appears as

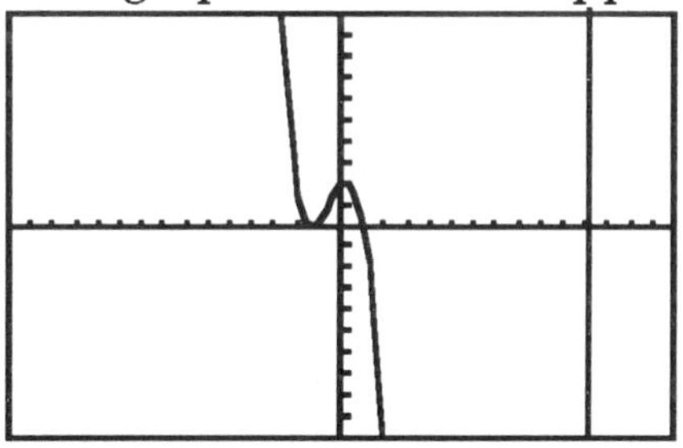

We now see the fourth intercept which appears to be between 12 and 13. Let's zoom in on this point. This we will find to be somewhat tricky, but it illustrates some of the limitations of the graphics calculator. If we evaluate the polynomial at 12 we find p = -7722 and at 13 we find p = 14616. You see here the incredible variation in p for x values between 12 and 13. With this in mind set the range to be [12, 13] x [-5000, 5000].

CASIO

On the older CASIO models, fx-7000G, fx7500G, fx-8000G, and fx-8500G enter on the screen or better yet in a program the commands

Graph Y=10Xx^{y}4-109Xx^{y}3-196X^2+109X+186 ◢
Factor 10

Execute these commands and when the graph is drawn activate the trace. Move the flashing dot to the point where the graph crosses the x-axis, approximately **X = 12.39361702** and press [EXE] twice. You will see the graph redrawn, but you have zoomed in by a factor of 10. Doing this twice more you will find **X = 12.4** is the intercept. On the fx-7700G set the zoom factors to be 10, ([F2] (**Zoom**) followed by [F2] (**FCT**) and enter 10 for each) set the range to be [12, 13] x [-5000, 5000] and redraw the graph. Activate the trace and move the cross hairs to the point where the graph crosses the x-axis, approximately **X = 12.393617**. Zoom in on this point ([F2] (**Zoom**) followed by [F3] (**xf**)). Trace to the point again you should find **X = 12.4** to be the intercept.

TI-81/82
Set the zoom factors to 10 and with the range set to [12, 13] x [-5000, 5000] draw the graph and from the ZOOM menu enter Zoom In. Move the cross hairs to the point between 12 and 13 where the graph crosses the x-axis, approximately X = 12.404255, and press [ENTER]. The graph is redrawn, but zoomed in by a factor of 10. Activating the trace we find that X = 12.4 is the intercept.

So the factors of the polynomial are $x + 1$, $x - 1$, and $2x - 3$. The Factor Theorem tells us that $x - 12.4$ is a factor as well. As a rational number, $12.4 = \frac{62}{5}$. As with the factor $2x - 3$, we will take the fourth factor to be $5x - 62$. Therefore we have

$$p = (x + 1)(x - 1)(2x - 3)(5x - 62).$$

The range settings for this example were arrived at guessing and evaluating the polynomial at a few points. This can be quite time consuming so to facilitate the drawing of the graphs the following programs will set **Ymin** and **Ymax** given **Xmin** and **Xmax** for a given function.

CASIO
A brief synopsis of the keys used to handle loops and logical tests will first be given.

Keys	Keystrokes	Position and Meaning
Implication Arrow	[SHIFT] [⇒]	Found above the [7] key. Used in conjunction with the relational operations, =, ≠, ≥, ≤, >, <, and other expressions:
fx-7700G	[SHIFT] [PRGM] [F1] (JMP) [F1] (⇒)	Examples: **A≥2⇒Goto3**
	A<0⇒4→X	The logic used here is a skipping idea. If the condition is true the command following the ⇒ is executed; otherwise the command is skipped.
Increment	[SHIFT] [Isz]	Found above the [4] and [1] keys,

fx-7700G [SHIFT] [PRGM] [F1] (JMP) [F5] (Dsz)	respectively. **Dsz** (decrease by 1 and skip
Decrement [SHIFT] [Dsz]	next command if zero) and **Isz** (increase by
fx-7700G [SHIFT] [PRGM]	1 and skip next command if zero) must be
[F1] (JMP) [F4] (Isz)	followed a letter of a memory location. If increments or decrements of values other than 1 are needed the implication arrow must be employed. Example: **0→I:Lbl 1** **. . . commands . . .** **I+2→I** **I≤10⇒Goto 1.**
Subprogram [Prog] (number)	This can be entered as a command in any
fx-7700G [SHIFT] [PRGM]	program. This will execute any other
[F3] (Prg)	program and when that program is finished the next line in the calling program will be executed.

Program 3.1.1C: Automatic Y Scaling

Keystrokes	Explanation
"A"→A:"B"→B ↵	Prompts for the input of **A** (**Xmin**) and **B** (**Xmax**).
(B-A)÷23→H ↵	Computes the spacing for the points at which the function is evaluated. Searches for the maximum and minimum of the function on 23 equally spaced points between **A** and **B**. The number 23 is chosen to lessen the likelihood of evaluating a function at a point where it is not defined.
A→X:Prog 0 ↵	Sets **X** equal to **A** and evaluates the function. This value becomes the initial value for **Ymin** and **Ymax**, which are stored in **C** and **D**. Function evaluation is done in program location 0.
Y→C:Y→D ↵	
0→I ↵	
Lbl 1 ↵	Sets label 1, the beginning of the loop.
A+IH→X ↵	Computes the **X** value.
Prog 0 ↵	Evaluates the function. Value stored in **Y**.
Y≤C⇒Y→C ↵	Replace **C** with **Y** if **Y** is smaller than **C**.

Y≥D⇒Y→D ↵	Replace **D** with **Y** if **Y** is larger than **C**
Isz I ↵	
I≤23⇒Goto 1 ↵	Continue the loop if **I** ≤ 23.
(D–C)/20→K	
Range A,B,(B–A)÷5,C–K,D+K, (D–C)÷5 ↵	Set the range.
Prog 1	Program location 1 contains the commands that draw the graph of the function.

TI-81

A brief synopsis of the commands used to handle loops, logical tests, and subprograms will first be given.

Key or Command	Keystroke	Position and Meaning
If	[PRGM] [▼] [▼] [ENTER]	Found in the **CTL** menu when on line 2 or greater of the edit screen for a program. Used in conjunction with the relational operators =, ≠, >, ≥. <, and ≤. Must be immediately followed on the next line with the command to be executed. Examples: **:If A<0** **:4→X** **:If A≥2** **:Goto 3**
TI-82	[PRGM] [ENTER]	
Increment	[PRGM] [▼] [▼] [▼] [ENTER]	Found in the **CTL** menu when on line 2 or greater of the edit screen for a program. These commands have the following form and meaning: **IS>(I, 10)** increase **I** by 1 and skip next command if **I** is greater than 10.
TI-82	Item **A** in **PRGM** menu	
Decrement	[PRGM] [▼] [▼] [▼] [▼] [ENTER]	**DS<(I, 1)** decrease **I** by one and skip next command if **I** is less than 1. If increments or decrements of values other than 1 are needed then the If command must be employed.
TI-82	Item **B** in **PRGM** menu	

Example: **:0→I**
:Lbl 1
. . . commands . . .
:I+2→I
:If I≤10
:Goto 1

Subprogram **PRGM** ▶ ▶ This moves to the **EXEC** title in the **PRGM** menu. There will then be the listing of all of the programs. Select the program you desire to be executed within the program you are editing. When this line is encountered the program selected will be executed. When that program is finished the next line in the calling program will be executed.

Program 3.1.1T: AUTOSCL, Automatic Y scaling

The equivalent TI-82 commands are given in parentheses.

Commands	Explanation
:All-Off (FnOff)	Unselects all expressions in the Y variables list.
:Disp "A"	Prompts for the input of **A** and **B**.
:Input A	
:Disp "B"	
:Input B	
:A→Xmin	Sets **Xmin** equal to **A** and **Xmax** equal to **B**.
:B→Xmax	
:(B–A)/5→Xscl	**Xscl** is set.
:(B–A)/23→H	Computes the spacing for the points at which the function is evaluated. Searches for the maximum and minimum of the function on 23 equally spaced points between **A** and **B**. The number 23 is chosen to lessen the likelihood of evaluating a function at a point where it is not defined.
:A→X	Sets **X** equal to **A** and evaluates the function. This value becomes the initial value for **Ymin** and **Ymax**, which are stored in **C** and **D**. Function evaluation is done using the expression in **Y1**.
:Y1→C	
:C→D	
:1→I	
:Lbl 1	Sets label 1, the beginning of the loop.
:A+IH→X	Computes the **X** value.
:If Y1≤C	Replace **C** with **Y1** if **Y1** is less than **C**.
:Y1→C	
:If Y1≥D	Replace **D** with **Y1** if **Y1** is larger than **D**.
:Y1→D	
:IS>(I,23)	Continues the loop if **I** ≤ 23.

:Goto 1	
:(D–C)/20→K	
:C–K→Ymin	Sets **Ymin**.
:D+K→Ymax	Sets **Ymax**.
:(D–C)/5→Yscl	Sets **Yscl**.
:Y_1-On	Selects **Y_1**.
:DispGraph	Displays the graph.

Let's apply these programs with the polynomial of example 3.

CASIO

In program location 0 enter

10Xx^y4-109Xx^y3–196X^2+109X+186→Y

and in program location 1 enter

Graph Y=10Xx^y4-109Xx^y3–196X^2+109X+186.

Execute the program and enter -3 for **A** and 15 for **B**. When the graph is drawn it should look like figure below left. If the program is reexecuted with -2 for **A** and 1.5 for **B** the graph will be like the figure below right.

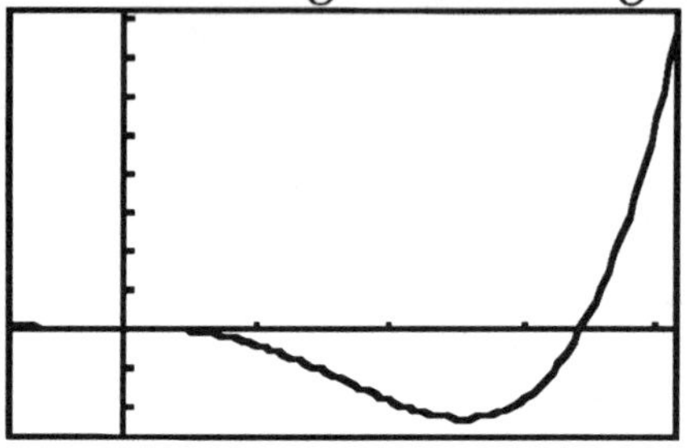

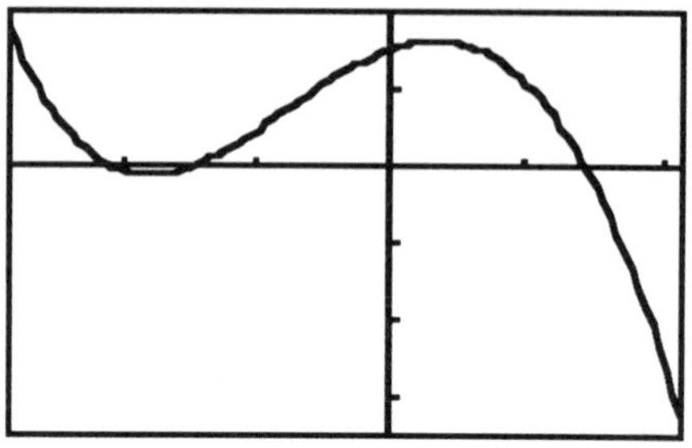

TI-81/82

For **Y_1** enter the expression **10X^4–109X^3–196X^2+109X+186**. Execute the program and enter -3 for **A** and 15 for **B**. The graph drawn should look like the figure above left. If we reexecute the program with -2 for **A** and 1.5 for **B** the graph will look like the figure above right.

3.1 Exercises

Factor each polynomial using the graph and the Factor Theorem?

1. $x^2 + 3x - 28$.
2. $5x^2 + 5x - 30$.
3. $2x^3 + x^2 - 8x - 4$.
4. $x^3 - 2x^2 - 9x + 18$.
5. $x^3 - 4x^2 - 11x + 30$.
6. $4x^3 + 22x^2 + 10x - 36$.
7. $2x^4 - 31x^3 + 86x^2 - 21x - 36$.
8. $6x^4 + 43x^3 - 201x^2 - 298x + 120$.

3.2 Solving Equations

(McKeague: Sections 3.9, 4.6, 4.7)

In the previous section we used a graph of a polynomial to find its factors. The Factor Theorem tells that to factor a polynomial p we must first find the values of x where p = 0, i.e., solve the equation p = 0. In this section we will build on this idea using a graph to solve other types of equations.

• **Example 1:** Solve the equation $3 + \frac{1}{x} = \frac{10}{x^2}$. (McKeague: Example 5, section 4.6)

As with the linear equations of chapter 2, we are search for points where the graphs of $y = 3 + \frac{1}{x}$ and $y = \frac{10}{x^2}$ meet. Set the range to the default setting and draw the graphs of these expressions.

CASIO: **Graph Y=3+1÷X:Graph Y=10÷X²**.

TI-81/82: **Y1=3+1/X; Y2=10/X²**.

The graph should appear as shown in the figure below. Activating the trace and moving

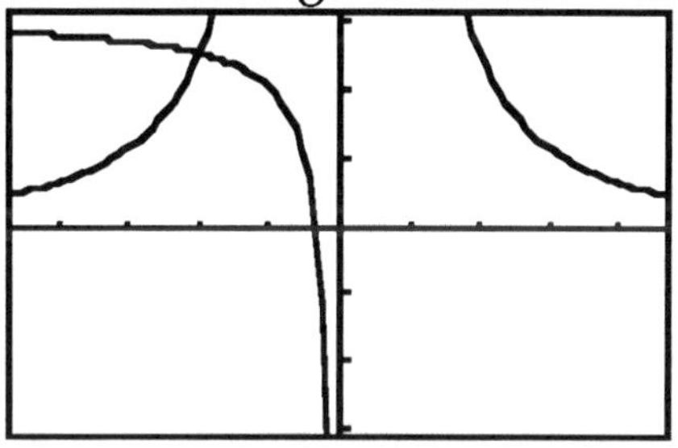

to the point where the graphs cross we find **X = -2**. Checking we find

$$3 + \frac{1}{-2} = 3 - \frac{1}{2} = \frac{5}{2} \text{ and } \frac{10}{(-2)^2} = \frac{10}{4} = \frac{5}{2}.$$

So x = -2 is a solution. To determine if there is another solution, let's zoom out by a factor of 2. This will produce the graph shown below. We see that there is another solution.

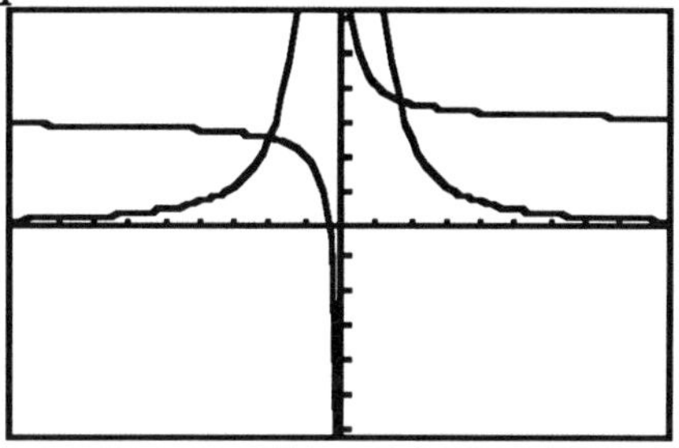

Again activating the trace we find an estimate of the solution to be **X = 1.6**. Zooming in on this point twice by a factor of 10 gives the estimate **X = 1.666**. Checking we find

$$3 + \frac{1}{1.666} = 3.600 \text{ and } \frac{10}{(1.666)^2} = 3.602$$

telling us that x = 1.666 is a good estimate for the solution. Zooming in twice more produces the estimate X = 1.66666. From this we infer that the solution is $x = \frac{5}{3}$. Checking we find that this is the exact solution.

The TI-82 has a built in utility for solving equations. It is the tenth item in the **MATH** menu. You will see it listed as **0:solve(**. The format of this command is **solve**(expression,variable,guess,{left,right}). The expression can have any single variable and that variable must be given as variable. Also you must have an expression where you are trying find where it equals zero. The value of guess is an intial estimate for the solution which can be obtained from a graph. The values of left and right provide a range of the utility to search for a solution and are optional. In the previous example where **Y1=3+1/X** and **Y2=10/X²** we could enter **solve(Y1-Y2,X,1.6)**. The calculator will return the value **1.666666667**.

3.2 Exercises

Find the solutions of each equation graphically and algebraically.

1. $\dfrac{2x-1}{x+4} = 1.$
2. $\dfrac{3}{x+3} + \dfrac{4}{x-5} = \dfrac{5}{4}.$
3. $\dfrac{x}{2} + \dfrac{3}{x} = \dfrac{7}{2}.$

3.3 Synthetic Division

(McKeague: Appendix)

In finding solutions of a polynomial equations the two theorems, Remainder Theorem and the Factor Theorem are used. With the graph of a polynomial and synthetic division it is possible to determine the real solutions of a polynomial equations. We first briefly review synthetic division and then give programs that carry out the procedure.

- <u>Example 2:</u> Use synthetic division to show that $x = 2$ is a root of the polynomial $p(x) = 2x^4 + 5x^3 - 14x^2 - 23x + 30$.

Synthetic division can be treated as a method for evaluating the polynomial, since by the Remainder Theorem, the remainder after dividing by $x - c$ is $p(c)$. to see how this treatment arises we write the polynomial in what is called the nested form. This nested form for p(x) is given by
$p(x) = (((2x + 5)x - 14)x - 23)x + 30.$
Taking $x = 2$ and performing the calculations as written, from the inside out, is precisely what is done in synthetic division. Let's carry out the calculations:

$$p(2) = (((\underbrace{2(2) + 5}_{})2 - 14)2 - 23)2 + 30$$
$$\parallel$$
$$((\underbrace{9(2) - 14}_{})2 - 23)2 + 30$$
$$\parallel$$
$$(\underbrace{4(2) - 23}_{})2 + 30$$
$$\parallel$$
$$\underbrace{-15(2) + 30}_{}$$
$$\parallel$$
$$0.$$

Since p(2) = 0, x = 2 is a root and so x – 2 is a factor of p(x). The calculations indicated by the symbol $\underbrace{\qquad\quad}$ are precisely those done in synthetic division. Writing synthetic division in the usual fashion we have

$$\begin{array}{r|rrrrr} 2 & 2 & 5 & -14 & -23 & 30. \\ & & 4 & 18 & 8 & -30 \\ \hline & 2 & 9 & 4 & -15 & 0 \end{array}$$

This tells that the quotient is $2x^3 + 9x^2 + 4x - 15$. Notice that the coefficients of the quotient are the intermediate values found when evaluating the polynomial in the nested form.

This nested form is precisely what is needed to construct a program that performs synthetic division. Since we will be using some new ways to store and access the memory locations, a brief discussion of this precedes the programs.

CASIO

The CASIO graphics calculators have the ability to utilize indexed memory locations. There are two keys, the brackets **[** and **]**, which are the **ALPHA** versions of the keys **•** and **EXP**, respectively. These are used to index memory locations. The following table shows how this works.

Indexed Location	Actual Location
A[0]	**A**
A[1]	**B**
C[12]	**O**
L[8]	**T**

All 26 (28 on the fx-7700G) memory locations can be accessed by using **A[0]**, **A[1]**, . . . , **A[25]**. It is also possible to extend the memory so that more memory locations beyond **Z** are usable. To see the total number of memory locations being used press **MODE** **•** **EXE** (**SHIFT** **Defm** **EXE** on the fx-7700G), producing the screen

```
** Defm **

Program  :  936

Memory   :  26

5598  Bytes  Free
```

fx-7000G, fx-7500G, fx-8000G, fx-8500G

```
Program     :   2699
F-Memory    :    124
Memory      :     28
Stat(SD)    :      0
Stat(REG)   :      0
Matrix      :      0
 1405 Bytes Free
```

fx-7700G

This screen gives the amount of total memory used by the program locations **0** – **9**, (**0** - **9**, **r**, **θ** on the fx-7700G) the number of memory locations, and how much memory is available. To extend the memory to 4 more locations beyond **Z** press [MODE] [·] [4] [EXE] ([SHIFT] [Defm] [4] [EXE] on the fx-7700G). These new locations are accessed as **Z[1]**, **Z[2]**, **Z[3]**, and **Z[4]**.

Caution: **Extending the memory will limit the amount of memory available for programs. This is especially critical in the CASIO fx-7000G.**

It is this indexed memory that we will use in the program for synthetic division to store the coefficients of the quotient.

Program 3.3.1C: Synthetic Division

Keystrokes	Explanation
"N"?→N:"C"?→C ↵	Prompt for the input of the degree **N** and **C**.
N→D:"COEF":N ◢	Set **U** equal to **N**, display **COEF** and the degree. This prompts for the input of the **N**th degree coefficient.
?→A ↵	Input the **N**th degree coefficient.
Lbl 1 ↵	Sets label 1, this is where the synthetic division takes place.
A→F[D–1] ↵	Place **A**, the (**D–1**)th degree coefficient of the quotient, in memory location **F[D–1]**.
"COEF":D–1 ◢	Display **COEF** and the degree of the next coefficient to be entered.
?→B:AC+B→A ↵	Enter the coefficient in **B** and perform a synthetic division step.
Dsz D ↵	Decrease **D** by one. If **D** is not zero, there are more coefficients to enter.
Goto 1 ↵	
"REM":A ◢	Display **REM** (remainder) and its value contained in **A**.

"QUOT":N→D ↵

Display **QUOT** (quotient). The loop defined by label 2 will display the coefficients of the quotient in ascending order.

Lbl 2 ↵
F[D–1] ◢
Dsz D ↵
Goto 2

Let's apply this program to the polynomial

$$p = 2x^4 + 5x^3 - 14x^2 - 23x + 30$$

with **C** = 2. Execute the program and enter the degree **N** = 4 and **C** = 2. The program responds with the screen shown to the left.

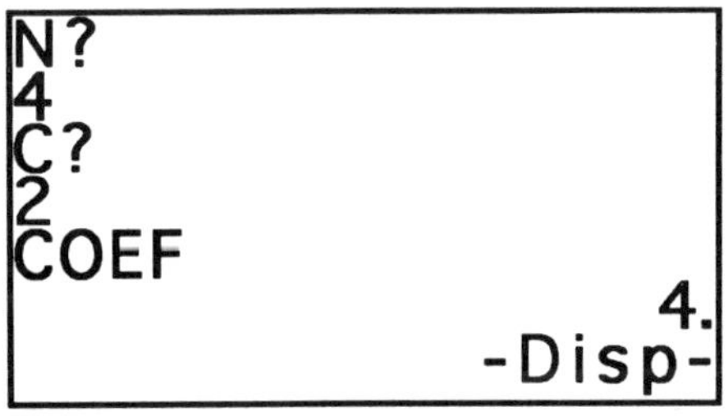

Press EXE and you are prompted to enter the coefficient of x^4, which is 2. Repeat this same procedure to enter in order the coefficients 5, -14, -23, and 30. 30 is the coefficient of x^0. The remainder (**REM**) is then displayed, which is zero, and pressing EXE four more times displays the coefficients of the quotient. We have then by the Factor Theorem

$$2x^4 + 5x^3 - 14x^2 - 23x + 30 = (x - 2)(2x^3 + 9x^2 + 4x - 15).$$

The coefficients of the quotient are stored in locations **F[3]**, **F[2]**, **F[1]**, and **F[0]** for later use. Now that we know one root of p, x = 2, and the coefficients of the quotient we can draw the graph of the quotient $q = 2x^3 + 9x^2 + 4x - 15$. See the figure below.

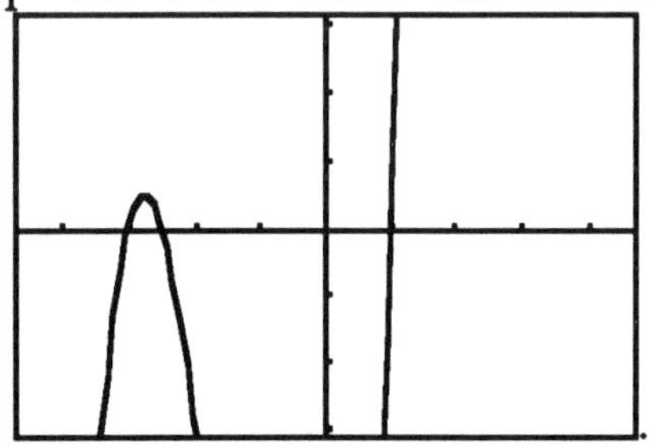

From this graph it appears that 1 is a root. Using the synthetic division program again and entering 3 for **N** and 1 for **C** and the coefficients 2, 9, 4, and -15 in that order gives a remainder of 0 and so we have that

$$2x^3 + 9x^2 + 4x - 15 = (x - 1)(2x^2 + 11x + 15).$$

Having now reduced to a quadratic polynomial we can simply factor $2x^2 + 11x + 15$ to see that the other roots are -2.5 and -3. Therefore

$$2x^4 + 5x^3 - 14x^2 - 23x + 30 = 2(x - 2)(x - 1)(x + 2.5)(x + 3).$$

The leading factor of 2 is needed since the leading coefficient of p is 2.

TI-81

The TI-81 has what is called **STAT** memory that we will use to store the coefficients of the quotient polynomial obtained in synthetic division. There are two parts to the **STAT** memory, **{x}** and **{y}**. This memory is indexed, so to access a value in either of these you must key in [2nd] [{x}] [1] [)] [ENTER]. This will display the first value held in the **STAT** memory **{x}**. If nothing is stored in this location, the preceding key sequence will cause a **MEMORY** error. To store an item in **{x}** enter the keystrokes [3] [STO▸] [2nd] [{x}] [1] [)] [ENTER]. This stores 3 in the first position of **{x}**. The entire contents of **STAT** memory can be viewed by first accessing the **STAT** menu, selecting the title **DATA**, and then selecting and entering the item **1: Edit**. All of the contents can be viewed by using the up and down arrows. Values can also be entered here by selecting the position, keying in the number and pressing [ENTER].

The following program performs synthetic division and makes use of the **STAT** memory **{x}** to store the coefficients of the quotient.

Program 3.3.1T81: SYNDIV Synthetic division

Commands	Explanation
:Disp "N" **:Input N** **:Disp "C"** **:Input C**	Prompts for the input of the degree **N** and **C**.
:1→I	
:Lbl 1 **:0→{x}(I)** **:IS>(I,N)** **:Goto 1**	This loop initializes the needed **STAT** memory locations. If this is not done a **MEMORY** error will occur.
:N→I	Set **I** equal to **N**.
:Disp "COEF" **:Disp N** **:Input A**	Display **COEF** and the degree. This prompts for the input of the **N**th degree coefficient which will be stored in **A**.
:Lbl 2	Label 2 sets the beginning of the loop where the synthetic division calculations take place.
:A→{x}(I)	Store the coefficient of the (**I–1**)th degree term in **{x}(I)**.
:Disp "COEF" **:I–1→J** **:Disp J**	Display **COEF**, set **J** equal to the degree of the coefficient to be entered and display its value.
:Input B	Input the degree (**I–1**)th coefficient.
:AC+B→A	Performs a synthetic division step.
:DS<(I,1)	Decrease **I** by 1. If **I** is not less than 1 there are more

:Goto 2	coefficients to enter.
:Disp "REM"	Display **REM** (remainder) and its value.
:Disp A	
:Pause	Pause. Press ENTER to continue.
:Disp "QUOT"	Display **QUOT** (quotient) and each of the
:N→I	coefficients of the quotient. This is the purpose of
:Lbl 3	the loop labelled by 3.
:Disp {x}(I)	
:DS<(I,1)	
:Goto 3	

TI-82

The TI-82 has the feature of being able to store and access lists. A list is a set of numbers enclosed in braces. The calculator can store up to 6 lists, **L1**, **L2**, **L3**, **L4**, **L5**, and **L6**. These are accessed from the 2nd function of the keys 1 – 6. To store a list enter **{1,2,5,7}** STO► 2nd L1 ENTER. To view the list enter 2nd L1 ENTER. To access one of the elements enter **L1(3)**. The value 5 will be displayed, i.e., the third element of the list **{1,2,5,7}**. We will use the list **L1** to store the coefficients of the quotient when performing synthetic division.

Program 3.3.1T82: SYNDIV Synthetic division

Commands	Explanation
:Disp "N"	Prompts for the input of the degree **N** and **C**.
:Input N	
:Disp "C"	
:Input C	
:N→dim L1	Sets the dimension of the list **L1**. The command **dim** is found in the **LIST** menu under the **OPS** title.
:Disp "COEF",N	Display **COEF** and the degree. This prompts for
:Input A	the input of the **N**th degree coefficient which will stored in **A**.
:N→I	
:Lbl 1	Label sets the beginning of the loop where the synthetic division takes place.
:A→L1(N–I+1)	Store the coefficient of (**N–I**)th degree term in correct location in **L1**.
:Disp "COEF",I–1	Display **COEF**, and the degree of the coefficient to
:Input B	be entered.
:AC+B→A	Perforems a synthetic division step.
:DS<(I,1)	Decrease **I** by 1. If I is not less than 1 there are more

:Goto 1	coefficients to enter.
:Disp "REM",A	Display **REM** (remainder) and its value.
:Pause	Pause. Press **ENTER** to continue.
:Disp "QUOT",L1	Display the list **L1** containing the coefficients in ascend order.

Let's apply this program to the polynomial

$$p = 2x^4 + 5x^3 - 14x^2 - 23x + 30$$

with **C** = 2. Execute the program and enter the degree **N** = 4 and **C** = 2. The program responds with the screen shown to the left. You are now prompted to enter the coefficient of x^4, which is 2. Enter this value and press **ENTER**. Repeat this procedure to enter in order the coefficients 5, -14, -23, and 30. 30 is the coefficient of x^0. The remainder (**REM**) is then displayed, which is zero, and pressing **ENTER** displays the coefficients of the quotient. The TI-82 will display **{2 9 4 -15}**. We have then by the Factor Theorem

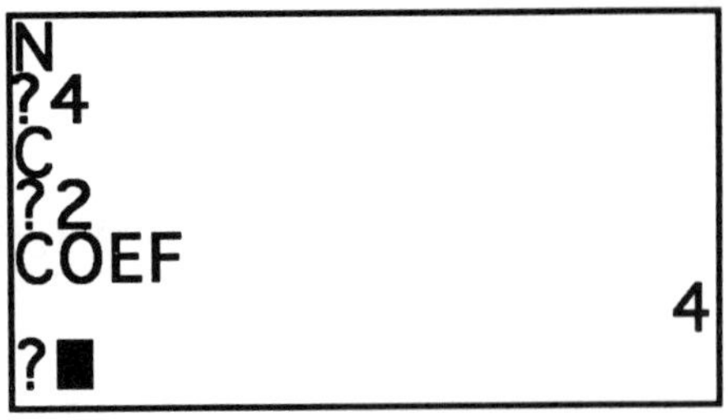

$$2x^4 + 5x^3 - 14x^2 - 23x + 30 = (x - 2)(2x^3 + 9x^2 + 4x - 15).$$

The coefficients of the quotient are stored in locations **{x}(4)**, **{x}(3)**, **{x}(2)**, and **{x}(1)** for later use. On the TI-82 the coefficients of the quotient are stored in **L1**. Now that we know one root of p, x = 2, and the coefficients of the quotient we can draw the graph of the quotient $q = 2x^3 + 9x^2 + 4x - 15$. See the figure below.

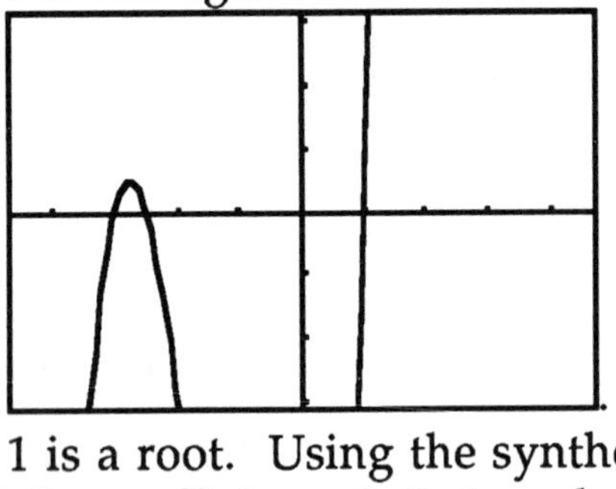

From this graph it appears that 1 is a root. Using the synthetic division program again and entering 3 for **N** and 1 for **C** and the coefficients 2, 9, 4, and -15 in that order gives a remainder of 0 and so we have that

$$2x^3 + 9x^2 + 4x - 15 = (x - 1)(2x^2 + 11x + 15).$$

Having now reduced to a quadratic polynomial we can simply factor $2x^2 + 11x + 15$ to see that the other roots are -2.5 and -3. Therefore

$$2x^4 + 5x^3 - 14x^2 - 23x + 30 = 2(x - 2)(x - 1)(x + 2.5)(x + 3).$$

The leading factor of 2 is needed since the leading coefficient of p is 2.

3.3 Exercises

For the polynomials given in the exercises of section 3.1 find the factorization using synthetic division.

For each of the polynomials find estimates for the real zeros. To check if you have an accurate estimate the remainder after synthetic division should be nearly zero.

1. $2x^3 + x^2 - 4x - 2$.
2. $3x^3 - 18.5x^2 - 4.5x - 45$.
3. $\frac{1}{4}x^4 - x^2 + \frac{1}{2}$.
4. $x^5 - 3x^3 + x + 1$.
5. $8x^4 + 22.8x^3 - 50.6x^2 - 94.8x + 138.6$.
6. $32x^6 - 48x^4 + 18x^2 - 1$.

7. In the development of synthetic division we look at the nested form of a polynomial given by
$$p(x) = 2x^5 + 7x^4 - 57x^3 - 157x^2 + 55x + 150$$
$$= ((((2x + 7)x - 57)x - 157)x + 55)x + 150.$$
Draw the graph of p(x) using both forms. Is there any difference in the time taken to draw the graphs? Explain this difference, if any.

9. As mentioned at the beginning of this chapter, polynomials are used for data fitting. The following table gives the population of the United States (in millions) every 10 years starting from 1900.

Year	1900	1910	1920	1930	1940	1950	1960	1970	1980
Population	75.995	91.972	105.711	123.203	131.669	150.697	179.323	203.212	226.505

Let x = 0 represent the year 1900, x = 1 represent the year 1910, etc. The following polynomials fit various values in the above table.

$p_1(x) = 18.81375x + 75.995$

$p_2(x) = 1.2238125x^2 + 9.02325x + 75.995$

$p_3(x) = 0.0342292x^3 + 0.317x^2 + 14.0870833x + 75.995$

$p_5(x) = -0.0373345x^5 + 0.6928452x^4 - 4.1395298x^3 + 8.9240119x^2 + 10.5370071x + 75.995$

$p_6(x) = -.0212702x^6 + 0.4731512x^5 - 3.8589857x^4 + 14.2379560x^3 - 23.2578583x^2 + 28.4040071x + 75.995$

$p_8(x) = -0.0049501x^8 + 0.1711748x^7 - 2.4224535x^6 + 17.9961278x^5 - 74.7758628x^4 + 171.0235236x^3 - 195.7667335x^2 + 99.7551738x + 75.995$

For example $p_2(0) = 75.995$, $p_2(4) = 150.697$, and $p_2(8) = 226.505$.

Plot the data in the table and on the same graph draw graphs of these polynomials. Comment on their accuracy in predicting the population in the years 1990 (x = 9) and 2000 (x = 10). Draw the graphs of these polynomials by entering them in the nested form used by Horner's Method, e.g.,

$p_5(x) = ((((-0.0373345x + 0.6928452)x - 4.1395289)x + 8.9240119)x + 10.5370071)x + 75.995.$

Investigating Areas Given a Fixed Perimeter

A rectangular area is to be fenced in with a perimeter P, length l, and width w having an area A.

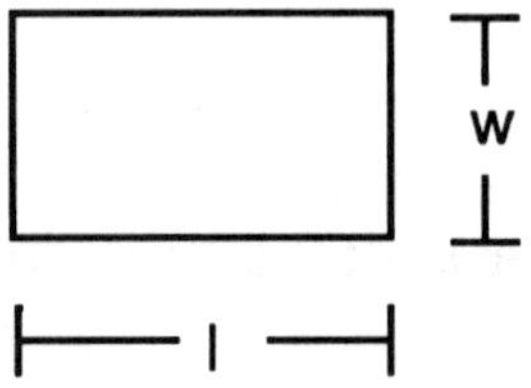

$P = 2w + 2l$ $\qquad$ $A = lw$

For a given value of P investigate the dimensions of the rectangle that has maximum area. Complete Table 1 to assist you in the investigation. The strategy for completing the table for the perimeter of 30 is illustrated below.

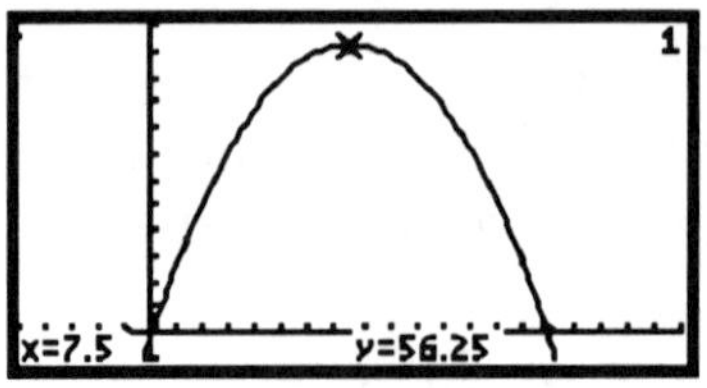

Figure 1

Figure 1 is from a TI-82 graphing calculator. The horizontal scale represents the width and ranges from -5 to 20, the vertical scale represents the area and ranges from -5 to 60. The maximum area with corresponding width is read from the graph using **TRACE**. Other model graphing calculators will give similar results.

P	$A = w\left(\frac{P}{2} - w\right)$	Maximum Area	Width	Length
30				
20				
40				
50				

Table 1

What is your observation concerning maximum area relative to the relationship between length and width for a given perimeter? Express this length and width in terms of the perimeter.

Now consider a slightly different situation. You have a rectangular area but it is located on a river so your fenced area has a perimeter P but it has only three sides, the river bank does not need to be fenced. Complete Table 2 to assist you in the investigation. For a given value of P investigate the dimensions of the rectangle that create maximum area.

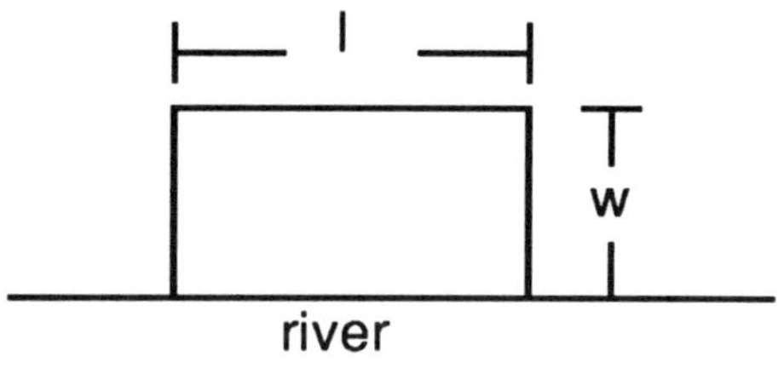

$P = 2w + l$ $\qquad$ $A = lw$

P	A = w (P – 2w)	Maximum Area	Width	Length
30				
20				
40				
50				

Table 2

What is your observation concerning maximum area relative to the relationship between length and width for a given perimeter? Express this length and width in terms of the perimeter.

Investigating Rational Exponents and Roots

1. Using your calculator evaluate $\left(\sqrt[3]{27}\right)^2$. If you entered 27^2/3, then the calculator answer is 243. However, the correct answer is 9. Investigate the correct way to enter the expression into the calculator by doing the following two problems as well as others you design.

   A. $64^{2/3}$ B. $\left(\frac{1}{8}\right)^{-4/3}$

If the base is positive, explain how to enter rational exponents into the calculator to get the correct answers.

2. Evaluate $\left(\sqrt[3]{-8}\right)^2$. If you entered -8^(2 ÷ 3), then the calculator answer is -4. This is a correct answer for the problem $-\left(\sqrt[3]{8}\right)^2$, but this is not the problem we set out to evaluate. When raising a negative base to a power the base **must** be put in parentheses, so enter $\left(\sqrt[3]{-8}\right)^2$ as (-8)^(2 ÷ 3). Depending on your calculator the answer will be error or a complex number, but the correct answer is 4. The calculator does not handle the rational exponent as you would expect.

   To investigate how the calculator processes rational exponents examine $\left(\sqrt[3]{-8}\right)^2$ by entering the following expressions into your calculator.

   A. (-8)^2^(1/3) B. (-8)^(1/3)^2

   C. (-8)^(1/3)^2 D. (-8)^2^(1/3)

Discuss the results of each of the above calculations. If the base is negative, explain how to enter rational exponents to insure correct and consistent results.

3. Think about the graph resulting from raising any number to the 2/3rds-power. How would you enter this function into the calculator? Two possible methods are shown below.

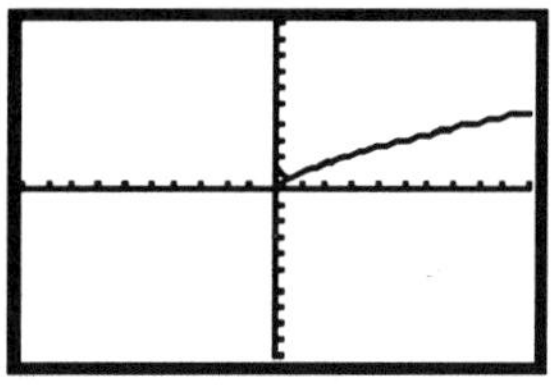

f(x) = x^(2/3)

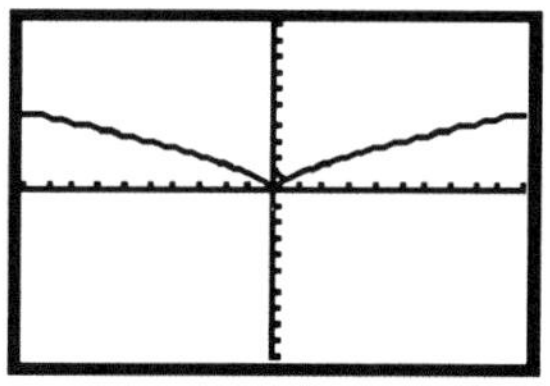

g(x) = x^(1/3)^2

Find the domain of f(x) and g(x). Discuss which graph best describes $y = x^{2/3}$.

Investigating the Domain of Rational Expressions

1. Using your calculator simplify $\frac{0}{5}$ and $\frac{5}{0}$.

Discuss the results of your calculations.

2. The domain of a rational expression includes those real numbers for which it is defined. For example, $\frac{1}{x-3}$ has a domain of x = {Real numbers, x≠ 3}. The graph of $y = \frac{1}{x-3}$ is illustrated in Figure 1. (Note: Graphing calculators usually have two graphing modes, one that plots just the calculated data points as in Figure 1(b) and one that connects the points as in Figure 1(a). When connecting the points the graph will appear to include a vertical line which is not actually part of the graph. In Figure 1(a) the vertical line is a vertical asymptote.)

(a) (b)

Figure 1

Investigate the domain of the following rational expressions by setting each one equal to y and graphing. Complete Table 1 to assist in your investigation. The domain of the first one has been completed for you.

Rational Expression	Domain
$\frac{1}{3x+12}$	x = {Real numbers, x ≠ -4}
$\frac{1}{x^2+5x+6}$	
$\frac{x-2}{2x-5}$	

Table 1

Discuss how to determine the domain of a rational expression.

3. The rational expression $\frac{x-3}{x^2-9}$ is not defined when x = +3 or x = -3 because at those values the numerical value of the expression is $\frac{0}{0}$ or $\frac{-6}{0}$. Yet when you graph the expression as shown in Figure 2 it appears to be defined at x = 3.

(a) (b)

Figure 2

Investigate the domain of the following rational expressions by setting each denominator equal to zero, thus finding the values for which it is undefined. Then graph each rational expression by setting it equal to y. Complete Table 2 to assist in your investigation. The actual and apparent domains for the first expression have been completed for you.

Rational Expression	Actual Domain	Apparent Domain from Graph
$\frac{x-1}{x^2+3x-4}$	x = {Real numbers, x ≠ -4, 1}	x = {Real numbers, x ≠ -4}
$\frac{2x-5}{-2x^2+15x-25}$		
$\frac{x}{5x^3-125x}$		

Table 2

Discuss how to determine when the graph will appear to be defined at a value where it is not defined. Could you determine the domain of a rational expression by just exploring the graph resulting when it is set equal to y without access to it symbolically?

Chapter 4

Lines and Inequalities

The simplest of all the equations in two variables is the linear equation. It is the foremost equation used to initially help to mathematically explain how particular systems vary with the input. This chapter shows how to analyze these types of equations using a graphics calculator. Along the way we will also use the calculator to visualize the solution of linear inequalities in two variables.

4.1 Lines

(McKeague: Sections 5.2, 5.3, 5.4)

Given two points in the plane (x_1, y_1) and (x_2, y_2) the slope of the line joining these points is given by

$$m = \frac{y_2 - y_1}{x_2 - x_1}$$

provided $x_1 \neq x_2$. Using this we find that the equation of the line through these points is given by

$$y - y_1 = m(x - x_1).$$

This is known as the point-slope form of the equation. As you know, if your goal is to draw the graph of the line it is better to use the slope-intercept form

$$y = mx + b.$$

The slope-intercept form is easily obtained from the point-slope form by taking

$$b = -mx_1 + y_1.$$

Using these formulas and given two points we can find the equation of the line passing through the points and draw its graph. The following programs can be used to do just that. The two points are taken as input from the graphic screen.

CASIO

Program 4.1.1C: Equation of a Line

Keystrokes	Explanation
Plot ◢	Sets a point at the screen center to be moved. Program halts. The screen coordinate is automatically stored in memory locations **X** and **Y**.
X→S:Y→T ↵	Stores the first point in **S** and **T**.
Plot ◢	Sets a point at the screen center to be moved. Program halts.
X→U:Y→V ↵	Stores the second point in **U** and **V**.
U–S→D ↵	Computes the denominator for the slope and stores the value in **D**.

D≠0⇒Goto 1 ↵	If **D** ≠ 0 then the line is not vertical and so go to label 1.
"NO SLOPE" ◢	Here **D** = 0, so there is no slope, i.e., a vertical line.
Plot S,T ↵	These two lines plot the points and draw the vertical line joining the points.
Plot U,V:Line ↵	
Goto 2 ↵	Go to label 2, the end of the program.
Lbl 1 ↵	Here, **D** ≠ 0 so the line is not vertical.
(V−T)÷D→M ↵	Compute the slope and store in **M**.
"M":M ◢	Display the value of the slope.
-MS+T→B ↵	Compute the y-intercept and store in **B**.
"B":B ◢	Display the value of the y-intercept.
Graph Y=MX+B ↵	Draw the graph of the line.
Lbl 2	

TI-81/82

The equivalent TI-82 command is in parentheses.

Program 4.1.1T: LINE Equation of a Line

Commands	Explanation
:All-Off (FnOff)	Unselects all of the expressions in the Y variables list.
:Input	This activates the graphics screen with movable cross hairs. The screen coordinate is automatically stored in memory locations **X** and **Y**.
:X→S	Stores the first point in **S** and **T**.
:Y→T	
:PT-On(S,T)	Plots the first point.
:Input	This activates the graphics screen with movable cross hairs.
:X→U	Stores the second point in **U** and **V**.
:Y→V	
:PT-On(U,V)	Plots the second point.
:U−S→D	Computes the denominator for the slope and stores the value in **D**.
:If D≠0	If **D** ≠ 0 then the line is not vertical, so go to label 1.
:Goto 1	
:Disp "NO SLOPE"	Here **D** = 0, so there is no slope, i.e., a vertical line.
:Pause	Program pauses. Press **ENTER** to continue.
:Line(S,Ymin,S,Ymax)	This draws the vertical line.
:Stop	Program stops.

:Lbl 1	Here, **D** ≠ 0 so the line is not vertical.
:Disp "M"	
:(V-T)/D→M	Compute the slope and store in **M**.
:Disp M	Display the value of **M**.
:Disp "B"	
:-MS+T→B	Computes the y-intercept and store in **B**.
:Disp B	Display the value of **B**.
:Pause	Program pauses. Press **[ENTER]** to continue.
:DrawF MX+B	Draw the graph of the line.

Before using these programs the range must be set. The programs 2.1.1C or 2.1.1T in section 2.1 are useful for this.

- <u>Example 1:</u> Find the equation of the line passing through the points (-6, 8) and (10, 11).

Set the range to be

CASIO: [-47, 47] x [-31, 31], **Xscl** = **Yscl** = 10
TI-81: [-48, 47] x [-32, 31], **Xscl** = **Yscl** = 10
TI-82: [-47, 47] x [-31, 31], **Xscl** = **Yscl** = 10.

This can be accomplished with the programs 2.1.1C or 2.1.1T using **X** = 0, **Y** = 0, and **S** = 1. Execute the above program and move the dot or cross hairs to the first point (-6, 8) and press **[EXE]** or **[ENTER]**. Move the dot or crosshairs to the second point (10, 11) and press **[EXE]** or **[ENTER]**. The values for the slope and intercept are **M** = **.1875** = $\frac{3}{16}$ and **B** = **9.125** = $\frac{73}{8}$. So the equation of the line is $y = \frac{3}{16}x + \frac{73}{8}$ or 16y = 3x + 146. Press **[EXE]** or **[ENTER]** again and the graph will be drawn. It should look like

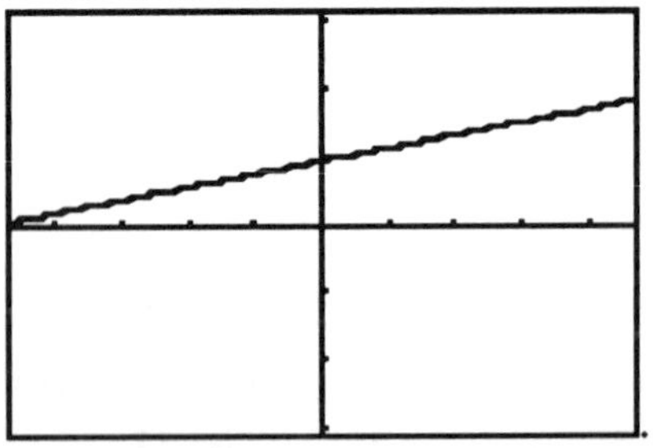

The programs above are easily changed to accept keyboard input. To do this simply make the following changes.

CASIO
Replace lines 1 through 4 with

```
"X1"?→S:"Y1"?→T ↵
"X2"?→U:"Y2"?→V ↵
```

TI-81
Replace lines 2 through 8 with

```
:Disp "X1"
:Input S
:Disp "Y1"
:Input T
:Disp "X2"
:Input U
:Disp "Y2"
:Input V
```

Similar changes can be made to accept from the keyboard the value of the slope and a point and then draw the graph.

- **Example 2:** Find the equation of the line that passes through (-3, 3) and (3, -1).

(McKeague: Example 5, section 5.3)

With the default range setting and using the above programs with the changes indicated Enter the points (-3, 3) and (3, -1) and we obtain **M = -.6666666667** = -2/3 and **B = 1**. The following graph is drawn.

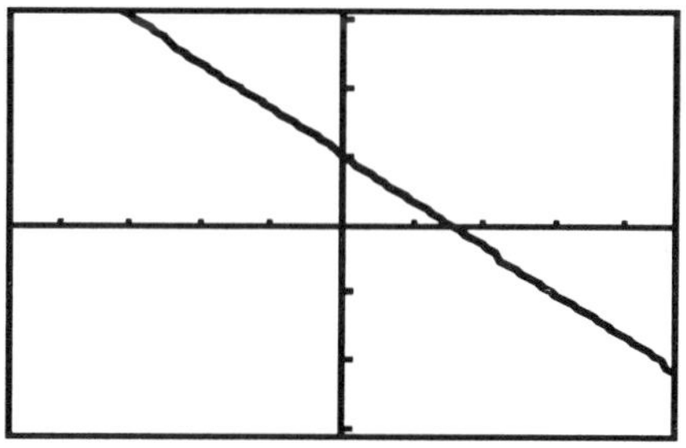

4.1 Exercises

Find the equation of the line with the given specifications and draw the graph.

1. Through the points (5, -1), (8, 6).
2. Through the points $\left(\frac{1}{3}, -8\right), \left(-5, \frac{1}{2}\right)$.
3. Slope m = -4, through (2, -1).
4. Slope $m = \frac{8}{5}$, through (10, 5).
5. Through the point (1, 3) and parallel to 2x + 3y = -4.
6. Through the point (-4, 0) and perpendicular to the line -4x + 5y = 12.

7. The price of a lawn trimmer varies seasonally from the manufacturer's suggested retail price (MSRP) of $85.00 on January 1 down to a spring sale price that is 25% off the MSRP for the month of May. On December 1 the price is back up to the MSRP. Determine the equations of the lines that describe the price of the lawn trimmer from January to May and from May to December.

4.2 Linear Inequalities

This section is devoted to the techniques used to determine the solution sets of systems of linear inequalities. Since different techniques are used for each calculator, we will discuss each separately. The general idea used throughout is the following.

The graph of y < f(x) consists of all points below the graph of y = f(x).

The graph of y > f(x) consists of all points above the graph of y = f(x).

CASIO

On the Casio it will require you to draw the graphs of the functions involved and then evaluate these functions at various points in the regions defined by the graphs. To this end, enter in program location 1 the commands that draw the graphs and in program location 0 the commands that evaluate the expressions defining the functions and display the values.

- <u>Example 1C:</u> Determine the solution set of the inequality $2x - 3y < 6$.

In program location 0 enter **"E":2X–3Y ◢** and in program location 1 enter

```
Graph  Y=2X÷3–2
Lbl 1:Plot ◢
Prog 0
Goto 1
```

Note that this program is an infinite loop, so [AC] must be pressed to halt the program.

Set the range to the default setting and execute program 1. The graph of y = 2x/3 – 2 is drawn and the flashing dot is at the origin. (It is a cross hairs if you are using the fx-7700G.) Move the dot to the point where **X = 1** and **Y = -1**. Remember to check both coordinates (**SHIFT** **X↔Y**) (this is not necessary on the fx-7700G) and record the ordered pair. The graph of this situation is given below on the left. Press **EXE** and the value of the left

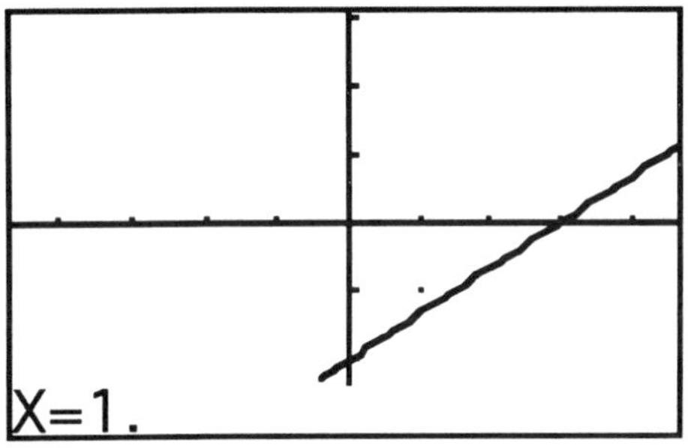

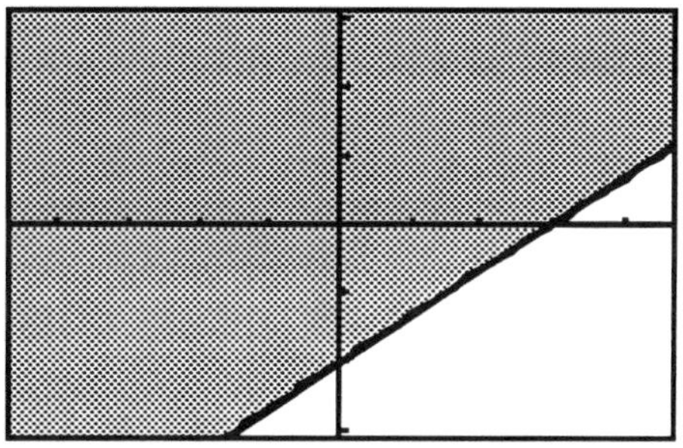

hand side of the inequality at (1, -1) is displayed; in this case the value is 5. So we find that the point (1, -1) satisfies the inequality. (The left hand side must be less than 6.) Move the dot to the point (3, -2). We find the value of the left hand side at this point to be 12. Since this value is greater than 6, we have found the region that defines the solution set. Note that the inequality can be written as y > 2x/3 – 2, so the solution set is the region above the graph of y = 2x/3 – 2. The graph of the solution set is depicted above on the right. The fx-7700G will draw the graph shown above on the right. First enter the **INEQ** graph type (**SHIFT** **MODE** **÷**) and enter **Graph** **F1** (Y>) **2X÷3–2** **EXE** .

- <u>Example 2C:</u> Determine the solution set of the system
  -5x + 10y > 3
  3x – 2y > -4

In program **0** enter

```
"E1":-5X+10Y ◢
"E2":3X–2Y ◢
```

and in program 1 enter

```
Graph  Y=X÷2+3÷10
Graph  Y=3X÷2+2
Lbl 1:Plot ◢
Prog 0:Goto 1
```

Set the range to be the default setting and clear the graphics screen. Execute program 1 and the graphs are drawn and the flashing dot (or crosshairs) is placed at the origin. The graph is shown below on the left. Move the dot to the point (-2, 1) and press **EXE**. We find

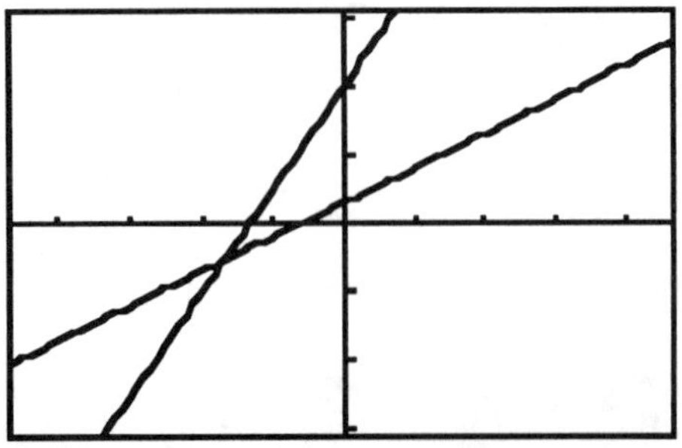

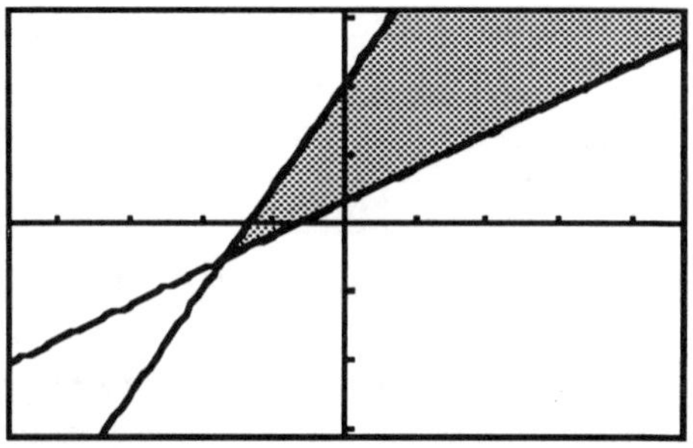

that **E1** is 20 and **E2** is -8. So (-2, 1) satisfies the first inequality but not the second. Now try the point (1, 3/2). In this case **E1** is 10 and **E2** is 0. So (1, 3/2) satisfies both inequalities. Try the point (-3, -2); **E1** is -5 and **E2** is -5. So we have discovered that the solution set is all the points above the line y = x/2 + 3/10 and below the line y = 3x/2 + 2. The solution set is depicted above on the right. With the fx-7700G the solution set can be drawn by first entering the **INEQ** graph type and then keying in

Graph Y>X÷2+3÷10
:Graph Y<3X÷2+2

The idea here is to write the inequalities in the form F(x, y) ≥ c (or >, ≤, or < where appropriate) where c is a constant. Then F(x, y) is the expression evaluated in program 0. The graphs are drawn by program 1 and points from the graph are input by using **Plot ◢**.

TI-81

The TI-81 has a shading utility that is very useful for drawing the solution sets of inequalities. The syntax for the shading command is

Shade(lower bound, upper bound, resolution, left bound, right bound**)**.

The value for the resolution, left bound, and right bound can be omitted. The defaults are respectively, **1**, **Xmin**, and **Xmax**. Below we give two examples.

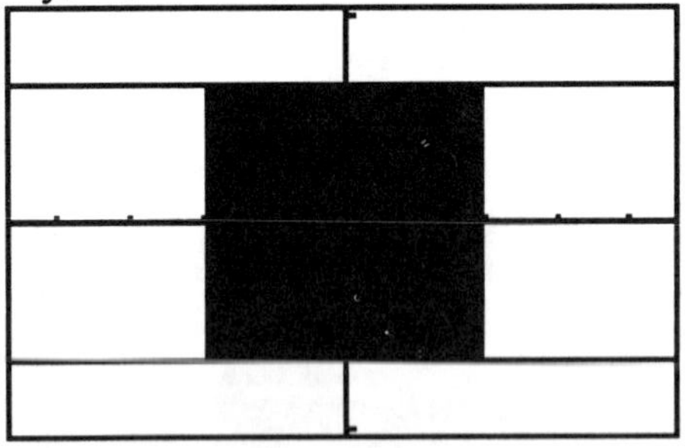

Shade(-2,2,1,-2,2)
TI-81: Range: [-4.8, 4.7] x [-3.2,3.1]
TI-82: Range: [-4.7, 4.7] x [-3.1,3.1]

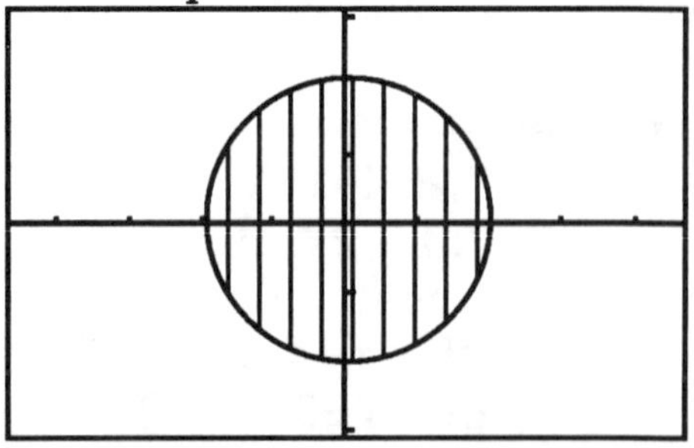

Shade($\sqrt{(4-x^2)}$,-$\sqrt{(4-x^2)}$,4)
TI-81: Range: [-4.8, 4.7] x [-3.2,3.1]
TI-82: Range: [-4.7, 4.7] x [-3.1,3.1]

- <u>Example 1T:</u> Determine the solution set of the inequality 2x – 3y < 6.

This inequality can be put in the form y > f(x) where f(x) = 2x/3 – 2. so the solution is all of the points above the graph of f(x) = 2x/3 – 2. So the command **Shade(2X/3–2,Ymax,1)** will draw the graph of the solution set. This is shown below.

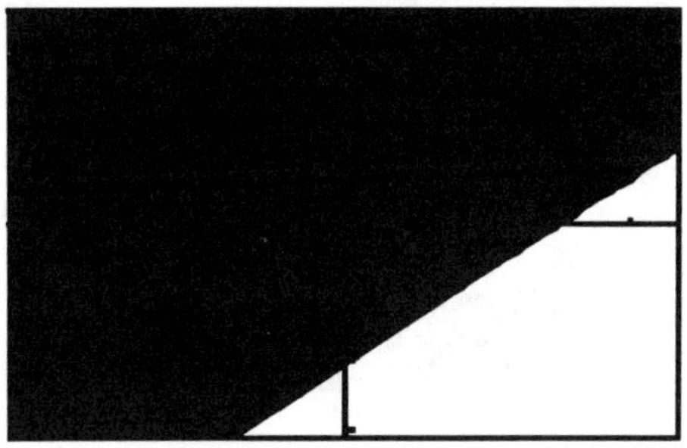

TI-81: Range: [-4.8, 4.7] x [-3.2,3.1]
TI-82: Range: [-4.7, 4.7] x [-3.1,3.1]

- **Example 2T:** Determine the solution set of the system
  $-5x + 10y > 3$
  $3x - 2y > -4$.

The first inequality can be written as $y > x/2 + 3/10$. So the solution of the first inequality is the set of points above the line $y = x/2 + 3/10$. The second inequality can be written as $y < 3x/2 + 2$. So its solution is the set of points below the line $y = 3x/2 + 2$. The commands needed to draw the graph of the solution are

Shade(X/2+3/10,Ymax,2)
Shade(Ymin,3X/2+2,4)

and these produce the graph

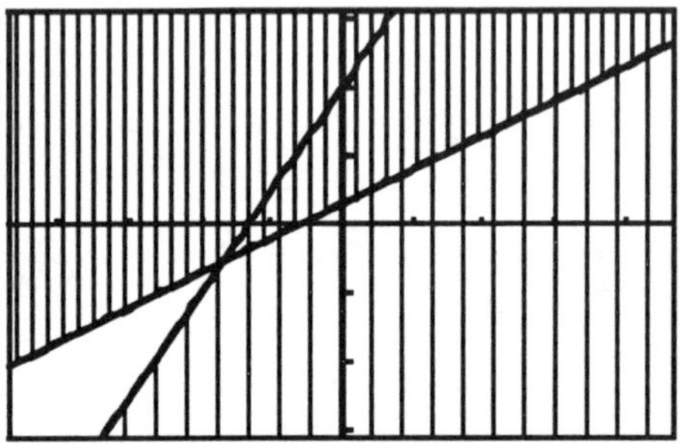

TI-81: Range: [-4.8, 4.7] x [-3.2,3.1]
TI-82: Range: [-4.7, 4.7] x [-3.1,3.1]

So the region common to both is below $y = 3x/2 + 2$ and above the line $y = x/2 + 3/10$.

It is sometimes easier to see the solution set if you draw the complement of each inequality, i.e., the region above the line $y = 3x/2+2$ and below the line $y = x/2 + 3/10$. The commands are

Shade(Ymin,X/2+3/10,1)
Shade(3X/2+2,Ymax,1)

and the graph will look like the figure at the top of the next page.

The following type of program is also helpful.

```
:DispGraph
:Lbl 1
:Input
:-5X+10Y→P
:3X−2Y→Q
:Disp "E1"
:Disp P
:Disp "E2"
:Disp Q
:Pause
:Goto 1
```

This program displays the graphs which are produced from the Y variables list. So **Y1=X/2+3/10** and **Y2=3X/2+2**. The range setting should be the default. After the graphs are drawn a crosshairs is placed at the screen center. Move the crosshairs to the point (-2, 1) and press **ENTER**. The values **E1** = 20 and **E2** = -8 are displayed. Since **E1** is the left hand side of the first inequality and **E2** is the left hand side of the second inequality we see that the point (-2, 1) satisfies the first inequality but not the second. Press **ENTER** again and you will return to the graph, ready to check more points. Move to the point (1, 3/2) and press **ENTER**. We find **E1** = 10 and **E2** = 0. So (1, 3/2) satisfies both inequalities. The solution set is then the set of points above the line $y = x/2 + 3/10$ and below the line $y = 3x/2 + 2$. Note that the above program has an infinite loop, so you must press **ON** to halt it.

The idea here is to draw the graphs of the functions from the Y variables list and then express each inequality in the form $F(x, y) \le c$ (or $<$, $\ge$, or $>$ where appropriate) where c is a constant. Then in the program evaluate each expression F(x, y) and display the value. In this way we test points from each region in order to find the solution set.

4.2 Exercises

Determine the solution set of each system of inequalities and draw their graph.

1. $2x - 3y > 1.$

2. $4x + 2y \leq 7.$

3. $y - x \leq 2.$
   $-y + 2x < 1$
   $2y + x > -4$

4. $2x - y \leq 1$
   $x + 3y < 4$
   $x \geq -3.$

5. $x + y \leq 6$
   $x - 6y \leq 12$
   $-3x + 2y \leq 12.$

Investigating Slope

The absolute value function, $f(x) = |x|$, is graphed in Figure 1. Suppose you graph the slope of the curve. The slope is -1 when $x < 0$ and +1 when $x > 0$. The slope has been graphed in Figure 2. Could you find an equation that would graph the slope? Graph $m(x) = \frac{|x|}{x}$ and compare your graph with the graph in Figure 2.

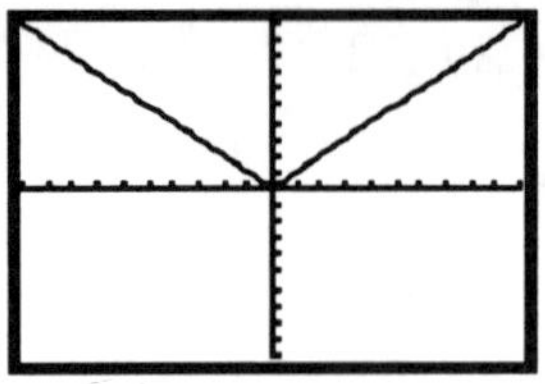

Figure 1

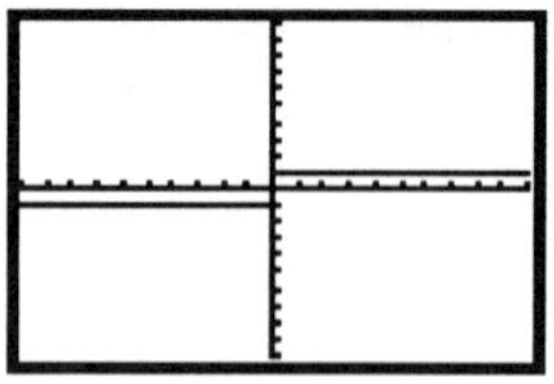

Figure 2

Investigate the equation that graphs the slope of a given absolute value by studying the following problem and completing Table 1. Figure 3 is the graph of the following equations.

$$f(x) = |x - 2|$$

$$m(x) = \frac{|x-2|}{x-2}$$

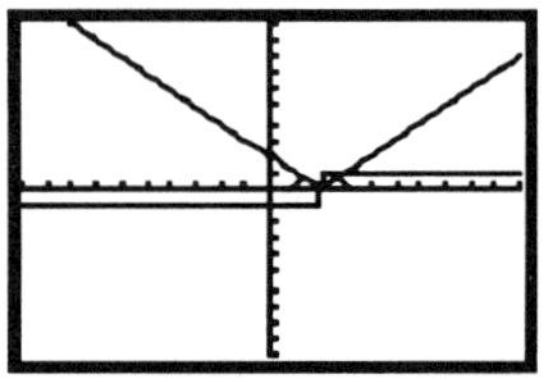

Figure 3

The "v-shaped" graph is the absolute value of x - 2 and the piece wise graph represents the slope of the absolute value of x - 2 for any value of x. Notice the slope of $y = x - 2|$ is -1 when $x < 2$ and +1 when $x > 2$.

Absolute Value Function	Point at Which Slope Changes	Slope to Left of Point	Slope to Right of Point	Equation That Graphs Slope of Absolute Value Function
$f(x) = \lvert x + 5 \rvert$				
$f(x) = \lvert 3x + 6 \rvert$				
$f(x) = \lvert 2x - 5 \rvert$				

Table 1

Based on your investigations and given $f(x) = \lvert ax + b \rvert$, *what is the equation that will graph the slope of f(x)?*

Chapter 5

Quadratic Equations

In this chapter we will investigate ways that the graphic calculator can aid in the solving of quadratic equations and inequalities. In the last chapter we learned how to graph functions and using tracing and zooming, found the x-intercepts of the functions. Building on these techniques we will learn about quadratic equations and inequalities.

5.1 Quadratic Equations

(McKeague: Sections 7.1, 7.2, 7.3, 7.4)

A quadratic equation is one that has the form

$$ax^2 + bx + c = 0$$

where a, b, and c are known real numbers with $a \neq 0$. Our goal here is to find all the values of x that, when substituted into the equation, make the left hand side of the equation zero. The solutions have a graphical interpretation also. Suppose we consider the function

$$f(x) = ax^2 + bx + c,$$

Then the solutions of the quadratic equation are the x-intercepts of the function. So the solutions can be found by graphing and zooming in on the x-intercepts. These were the techniques developed in Chapter 1. Rather than repeat that procedure, which can take some time, we will program the calculator to use the quadratic formula.

Quadratic Formula

The solutions of $ax^2 + bx + c = 0$ are given by

$$x = \frac{-b \pm \sqrt{b^2 - 4ac}}{2a}.$$

An important quantity in the formula above is the discriminant: $D = b^2 - 4ac$. It has the following meaning:

Discriminant	Solutions
$D > 0$	Two real solutions.
$D = 0$	One real solution.
$D < 0$	No real solutions.

Graphically the discriminant tells the number of x-intercepts of the quadratic function $f(x) = ax^2 + bx + c$. The following graphs illustrate this fact.

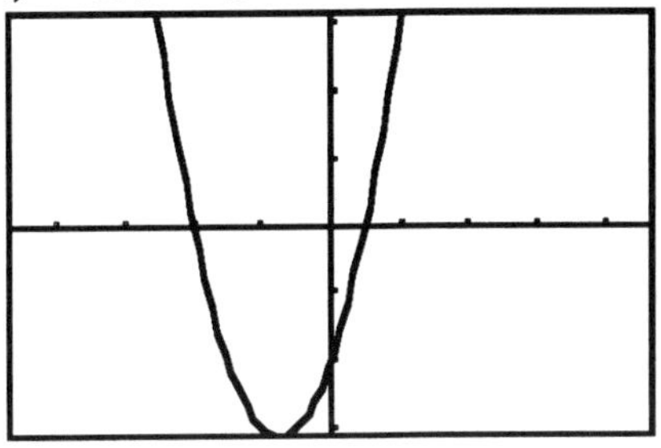

D > 0: $f(x) = 2x^2 + 3x - 2$

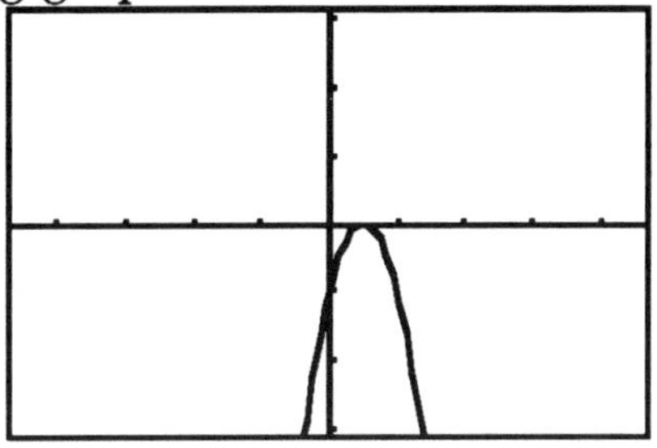

D = 0: $f(x) = -4x^2 + 4x - 1$

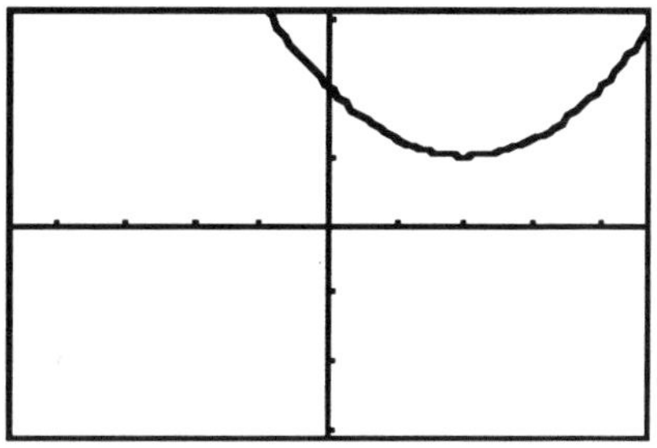

D < 0: $f(x) = \frac{1}{4}x^2 - x + 2$

In the case that D < 0 the solutions are complex numbers. These are numbers of the form $a + bi$ where a and b are real numbers and $i^2 = -1$. For example the solutions of

$$2x^2 - x + 2 = 0$$

are the complex numbers $\frac{1}{4} + \frac{\sqrt{15}}{4}i$ and $\frac{1}{4} - \frac{\sqrt{15}}{4}i$. For our needs when D < 0 we will give the real and imaginary parts of the complex solutions, i.e., for the solutions above, give only $\frac{1}{4}$ and $\frac{\sqrt{15}}{4}$.

In a program that utilizes the quadratic formula, there is a need to determine if the discriminant is positive, negative, or zero, i.e., the need for logical tests and jumping to the appropriate instructions. Let's look at how each calculator does this.

CASIO

We first give a synopsis of the necessary program commands used for logical testing.

Key	**Keystroke**	**Position and Meaning**
Lbl	[SHIFT] [Lbl] (number)	Found above the [⇐] key. Labels a line or section of commands and is followed by a number in the range 0-9. Used in conjunction with Goto.
fx-7700G	[SHIFT] [PRGM] [F1] (JMP) [F3] (Lbl)	
Goto	[SHIFT] [Goto] (number)	Found above the [Prog] key. Causes the program to jump to the line labeled with a number. Followed by a number in the range 0-9. Used in conjunction with Lbl.
fx-7700G	[SHIFT] [PRGM] [F1] (JMP) [F2] (Gto)	
⇒	[SHIFT] [➡]	Found above the [7] key. Used for logical tests. Uses the syntax: **(expression) (relational**
fx-7700G	[SHIFT] [PRGM]	

[F1] (JMP) [F1] (⇒) — operator: <, ≤, >, ≥, =, ≠) (expression) [➡] (command).

Relational Operators	[SHIFT] [>]	Found above the keys [2], [3], [5], [6], [8], and [9], respectively. Used for the logical tests given in the above command.
	[SHIFT] [<]	
	[SHIFT] [≥]	
	[SHIFT] [≤]	
	[SHIFT] [=]	
	[SHIFT] [≠]	
fx-7700G	[SHIFT] [PRGM] [F2] (REL)	

Let's put all of these programming commands together in a program that utilizes the quadratic formula to solve a quadratic equation. Put the following program in location 3.

Program 5.1.1C: Quadratic Formula

Keystrokes	Explanation
"A"?→A:"B"?→B ↵	Prompts for the input of the coefficients a and b. Recall the : concatenates lines.
"C"?→C ↵	Prompts for the input of the coefficient c.
"D":B²–4AC→D ◢	Computes and displays the discriminant.
D>0⇒Goto 1 ↵	Checks if the discriminant is positive. If it is positive, go to label 1, otherwise go to the next command.
D=0⇒Goto 2 ↵	Checks if the discriminant is zero. If it is zero, go to label 2, otherwise go to the next command.
D<0⇒Goto 3 ↵	Checks if the discriminant is negative. If it is negative, go to label 3.
Lbl 1:"ROOTS" ↵	Here the discriminant is positive so there are two roots.
(-B+√D)÷(2A)→R ◢	Computes one root, stores it in **R** and displays the value.
(-B–√D)÷(2A)→S ◢	Computes the other root, stores it in **S** and displays the value.
Goto 4 ↵	Jumps to label 4, which ends the program.
Lbl 2:"ROOT" ↵	Here the discriminant is zero so there is only one root.

-B÷(2A)→R ◢	Computes the one root, stores it in **R** and displays the value.
Goto 4 ↵	Jumps to label 4, which ends the program.
Lbl 3:"COMPLEX" ↵	Here the discriminant is negative so there are two complex roots.
"RE":-B÷(2A)→R ◢	Computes the real part, stores it in **R** and displays the value.
"IM":√-D÷(2A)→I ◢	Computes the imaginary part, stores it in I and displays the value. Roots are **R** ± **I***i*.
Lbl 4	End of program.

- **Example 1:** Using the quadratic formula find the solutions of the equations
  (a) $6x^2 + 7x - 5 = 0.$ (McKeague: Example 1, 7.2)
  (b) $\frac{1}{4}x^2 - x + 2 = 0.$

Run the above program by first entering the **RUN** mode ([MODE] [1]) and keying in [Prog] [3] [EXE] , or on the fx-7700g, [SHIFT] [PRGM] [F3] (**Prg**) [3] [EXE]. Enter the coefficients of the first equation **A** = 6, **B** = 7, and **C** = -5. The screen will look like the figure on the left below.

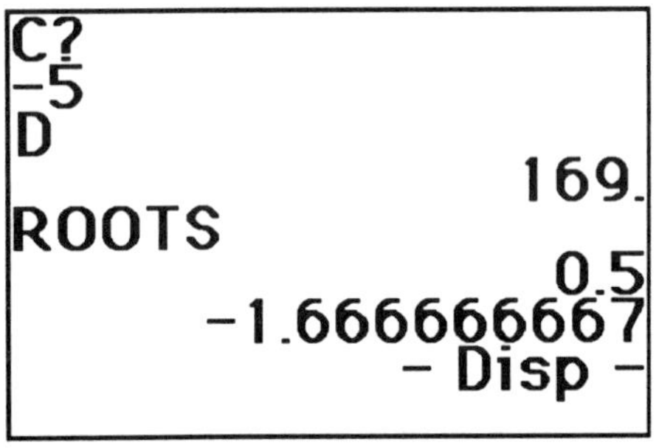

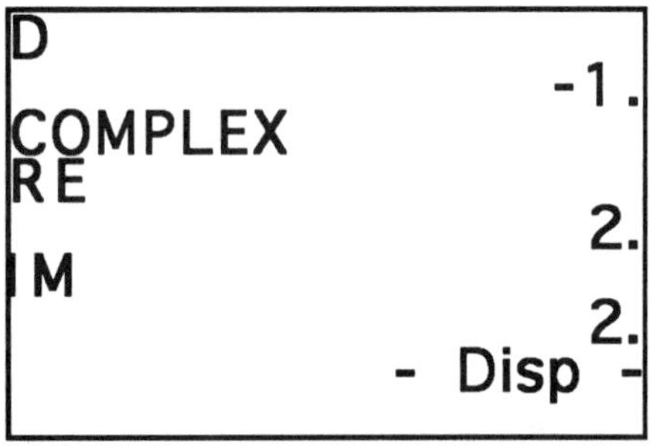

The solutions are then 0.5 = 1/2 and -1.666666667 = -5/6. Reexecute the program and enter the coefficients of the second equation **A** = .25, **B** = -1, and **C** = 2. The screen will appear as the figure on the right above. The solutions are the complex numbers 2 ± 2*i*.

TI-81

We first give a synopsis of the programming commands necessary for logical testing.

Key or Command	Keystroke	Position and Meaning
Lbl	[PRGM] [ENTER] **(number)**	Found in the **CTL** menu when on line 2 or greater of the edit screen for a program. Labels a line or section of commands and is followed by a number in the range 0-9. Used in conjunction with Goto.
TI-82	Item 9 in **CTL** menu	

Goto TI-82	PRGM ▼ ENTER (number) item 10 in **CTL** menu	Found in the **CTL** menu when on line 2 or greater of the edit screen for a program. Causes the program to jump to the line labeled with a number. Followed by a number in the range 0-9. Used in conjunction with Lbl.
If TI-82	PRGM ▼ ▼ ENTER PRGM ENTER	Found in the **CTL** menu when on line 2 or greater of the edit screen for a program. Used in conjunction with the relational operators. Uses the syntax: **If (expression) (relational operator: <, ≤, >, ≥, =, ≠) (expression).**
Relational Operators	2nd TEST	All found in the **TEST** menu screen. Used for the logical tests with the **If** command.
=	ENTER	
≠	▼ ENTER	
>	▼ ▼ ENTER	
≥	▼ ▼ ▼ ENTER	
<	▼ ▼ ▼ ▼ ENTER	
≤	▼ ▼ ▼ ▼ ▼ ENTER	

Let's put all of these commands together in a program that utilizes the quadratic formula to solve a quadratic equation. Name this program **QUADF**.

Program 2.1.1T: QUADF, Quadratic Formula

Commands	Explanation
:Disp "A"	Displays the letter **A**.
:Input A	Prompts for the input of the coefficient a.
:Disp "B"	Displays the letter **B**.
:Input B	Prompts for the input of the coefficient b.
:Disp "C"	Displays the letter **C**.
:Input C	Prompts for the input of the coefficient c.
:B^2–4AC→D	Computes the discriminant, stores it in D.
:Disp "D"	Displays the letter **D**.
:Disp D	Displays the value of the discriminant.

:If D>0	Checks if **D** is positive; if so go to label 1, if not skip the next instruction.
:Goto 1	
:If D=0	Checks if **D** is zero; if so go to label 2, if not skip the next instruction.
:Goto 2	
:If D<0	Checks if **D** is negative; if so go to label 3.
:Goto 3	
:Lbl 1	Here the discriminant is positive, so there are two roots.
:Disp "ROOTS"	Displays the word **ROOTS**.
:(-B+√D)/(2A)→R	Computes the first root.
:Disp R	Displays the value of the first root.
:(-B–√D)/(2A)→S	Computes the second root.
:Disp S	Displays the value of the second root.
:Goto 4	Jumps to label 4, which ends the program.
:Lbl 2	Here the discriminant is zero, so there is only one root.
:Disp "ROOT"	Display the word **ROOT**.
:-B/(2A)→R	Computes the root.
:Disp R	Displays the value of the root.
:Goto 4	Jumps to label 4, which ends the program.
:Lbl 3	Here the discriminant is negative, so there two complex roots.
:Disp "COMPLEX"	Displays the word **COMPLEX**.
:Disp "RE"	Displays the letters **RE**, meaning real part.
:-B/(2A)→R	Computes the real part of the roots.
:Disp R	Displays the value of the real part.
:Disp "IM"	Displays the letter **IM**, meaning imaginary part.
:√-D/(2A)→I	Computes the imaginary part.
:Disp I	Displays the value of the imaginary part. Roots are **R** ± **I***i*.
:Lbl 4	Ends the program.

- <u>Example 1:</u> Using the quadratic formula find the solutions of the equations
  (a) $6x^2 + 7x - 5 = 0$. (McKeague: Example 1, 7.2)
  (b) $\frac{1}{4}x^2 - x + 2 = 0$.

Run the above program by returning first to home screen (**2nd** **QUIT**) and then pressing **PRGM**. With the **EXEC** title selected, select the program number or letter containing **QUADF** and press **ENTER**. Enter the coefficients of the first equation **A** = 6,

B = 7, and **C** = -5. The screen will look like the figure below left. The solutions are

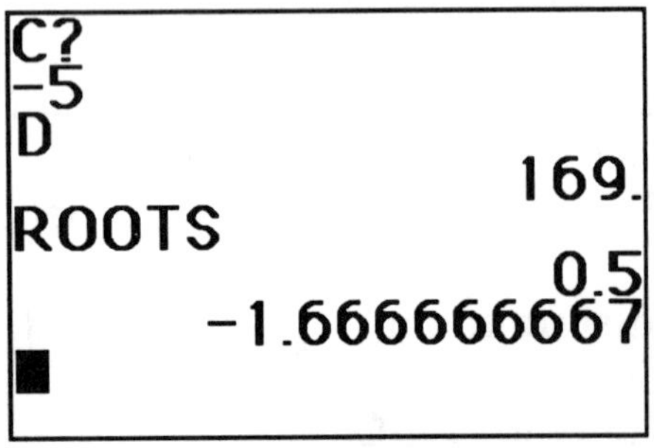

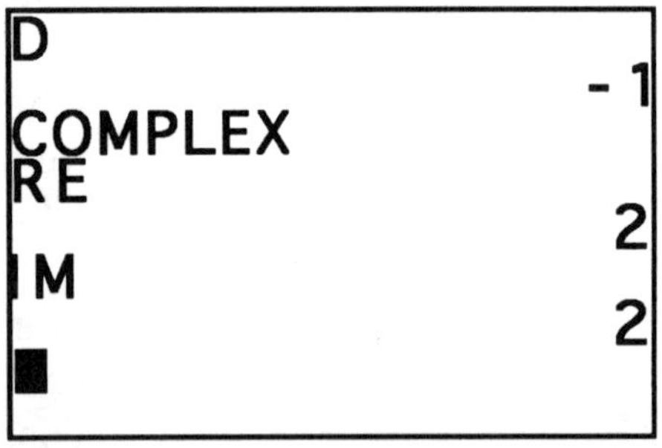

then 1.435781669 and -0.8357816692. Reexecute the program and enter the coefficients of the second equation **A** = .25, **B** = -1, and **C** = 2. The screen will appear as the figure above right. The solutions are the complex numbers $2 \pm 2i$.

- **Example 2:** Find the solutions of the equations
  (a) $3x^6 + 13x^3 - 10 = 0$.
  (b) $x + \sqrt{x} - 6 = 0$ (McKeague: Example 3, Section 7.4)

(a) If $u = x^3$ then $3x^6 + 13x^3 - 10 = 3u^2 + 13u - 10 = 0$. Using the quadratic formula program with **A** = 3, **B** = 13, and **C** = -10, the roots found are **R** = 0.6666666667 = 2/3 and **S** = -5. Since $u = x^3$, we require x^3 = **R** or $x = \sqrt[3]{2/3} = 0.8735804647$ and x^3 = **S** or $x = \sqrt[3]{-5} = -1.709975947$.

(b) If $u = \sqrt{x}$ then $x + \sqrt{x} - 6 = u^2 + u - 6 = 0$. Using the quadratic formula program with **A** = 1, **B** = 1, and **C** = -6, the roots found are **R** = 2 and **S** = -3. Since $u = \sqrt{x}$, we require $\sqrt{x}$ = **R** or $x = 2^2 = 4$ and $\sqrt{x}$ = **S** or $x = (-3)^2 = 9$. Checking these values we find that x = 4 is the only solution.

- **Example 3:** A football is kicked straight up from a height of 3.5 feet above a level playing field with an initial velocity of 62 feet per second. Find when the football hits the ground and how high it will go.
  It is known that if an object is thrown straight up or dropped straight down from a height h_0 with a velocity of v_0 then its height at any time t is given by the quadratic function

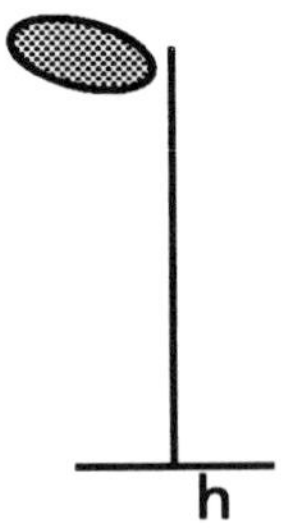

$$h = -16t^2 + v_0t + h_0.$$

In this situation the height of the football at any time t is given by the equation

$$h = -16t^2 + 62t + 3.5.$$

The football will hit the ground when h = 0, i.e., when $-16t^2 + 62t + 3.5 = 0$. Using the calculator we find two solutions to this equation, namely,

$$t_1 = -0.0556523399,\ t_2 = 3.93065234.$$

Since t represents time, the only solution that makes sense is t = 3.93065234. So approximately 3.9 seconds after being kicked the football will hit the ground. To find how high the football will go, draw the graph of the function

$$y = -16x^2 + 62x + 3.5.$$

Here x will represent time and y will be height. Using the range [0, 4] x [0, 65] we get the

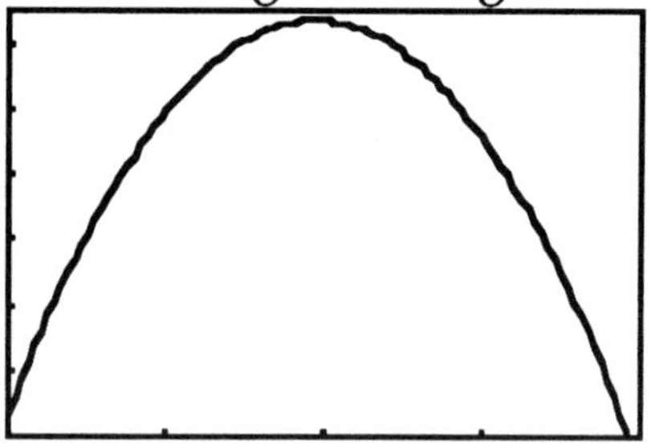

graph shown above. Using the trace we find that the highest point or maximum point on the graph is approximately **X = 1.96** and **Y = 63.6**. So after about 1.96 seconds the ball reaches its maximum height of approximately 63.6 feet.

• __Example 4:__ A projectile is shot from a height h_0 with an initial velocity of v_0 and an upward slope of m. See the figure below. (The slope is the

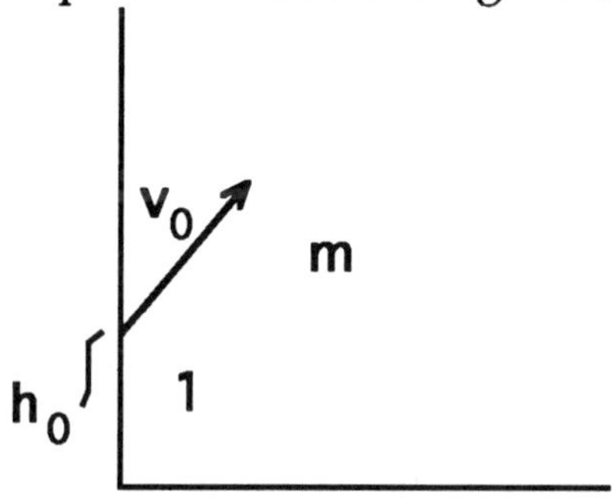

ratio of the vertical change to the horizontal change. If m = 1, the projectile is shot upward at an angle of 45°. For other values see the table that follows). If x is the horizontal distance travelled by the projectile then

angle	10°	20°	30°	45°	50°	60°	70°	80°	85°
m	0.176	0.364	0.577	1	1.192	1.732	2.747	5.671	11.430

the path of the object, neglecting wind resistance, is given by the graph of the quadratic function

$$y = \frac{-16\left(1 + m^2\right)}{v_0^2} x^2 + mx + h_0$$

where $x \geq 0$. Suppose $v_0 = 176$ feet per second, $h_0 = 3$ feet, and $m = 0.75$.

(a) Find where the projectile will hit the ground. That is, the horizontal distance travelled by the projectile.

(b) Find the maximum height of the projectile.

(c) Find the value of m that gives the maximum horizontal distance travelled.

The following programs will be helpful. The equivalent TI-82 commands are in parentheses.

CASIO	TI-8/82
"M"?→M	**:All-Off (FnOff)**
"V"?→V	**:Disp "M"**
"H"?→H	**:Input M**
-16(1+M²)÷V²→A	**:Disp "V"**
M→B	**:Input V**
H→C	**:Disp "H"**
Graph Y=AX²+BX+C	**:Input H**
	:-16(1+M²)/V²→A
	:M→B
	:H→C
	:DrawF AX²+BX+C

Recall the command **All-Off** is found in the **Y-VARS** menu under the **Off** title and unselects all of the functions stored in the Y variables list. On the TI-82 it **FnOff** is in the **Y-VARS** menu in the **On/Off** item. The **DrawF** command is found in the **DRAW** menu and causes the calculator to draw the graph of any expression in the variable **X**. This command allows you to have several graphs on the screen at once, not just the 4 stored in the Y variables list. Unfortunately the trace function cannot be used in conjunction with the **DrawF** command, but a moving crosshairs can be activated after the graph is drawn by pressing any of the arrow keys. The solution to Example 4 follows.

(a) Set the range of the calculator to be [0, 1000] x [0, 250] with the x and y scale values to be 100. Execute the above program, entering **M** = .75, **V** = 176, and **H** = 3. The graph will appear as

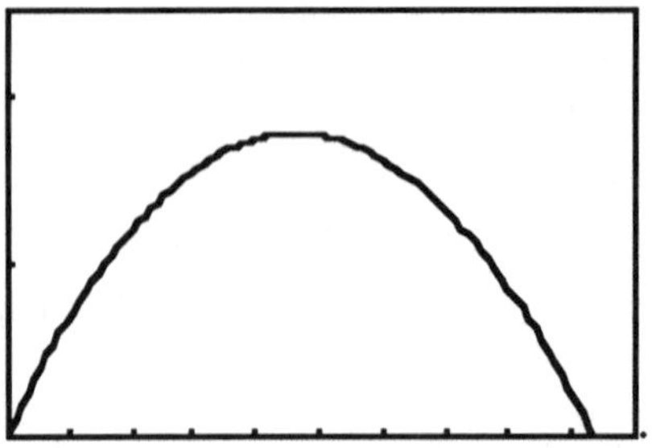

We see from the graph that the point where the projectile hits the ground is a solution of the equation

$$\mathbf{A}x^2 + \mathbf{B}x + \mathbf{C} = 0$$

where $\mathbf{A} = \frac{-16\left(1+\mathbf{M}^2\right)}{\mathbf{V}^2} = -8.070764463\text{x}10^{-4}$, **B** = .75, and **C** = 3. So using the quadratic formula program and entering **A, B,** and **C**, i.e., when prompted for **A, B,** and **C** enter [ALPHA] [A], [ALPHA] [B], and [ALPHA] [C]. (The appropriate values for these coefficients were stored in **A, B,** and **C** by the programs given above.) The solutions are -3.982929016 and 933.262929. Since x represents distance, the projectile travels a horizontal distance of about 933.26 feet.

(b) Reexecute the program given above with the same values of **M, V,** and **H**. Using the trace on the CASIO and moving to the maximum point of the graph, we find the highest point to be **Y = 177.230421**. With TI-81, enter **AX² + BX + C** into location **Y1** of the Y variables list and draw the graph. The trace is now active and we find the highest point to be **Y = 177.238233**. The projectile reaches a maximum height of about 177.2 feet.

(c) The value of x where the projectile hits the ground is the horizontal distance travelled and for $h_0 = 3$, $v_0 = 176$, and any value of m, x is the solution of the quadratic equation

$$\frac{-16\left(1+m^2\right)}{30976}x^2 + mx + 3 = 0$$

By the quadratic formula,

$$x = \frac{-m \pm \sqrt{m^2 - 4\left(\frac{-16(1+m^2)}{30976}\right)(3)}}{2\left(\frac{-16\,(1+m^2)}{30976}\right)}$$

$$= \frac{-968\left(-m \pm \sqrt{m^2 + \frac{3\,(1+m^2)}{484}}\right)}{1+m^2}$$

$$= \frac{-968\left(-m \pm \sqrt{\frac{487m^2+3}{484}}\right)}{1+m^2}$$

$$= \frac{968m \pm 44\sqrt{487m^2+3}}{1+m^2}$$

Choosing the positive sign so that x > 0 we want to find m so that

$$x = \frac{968m + 44\sqrt{487m^2+3}}{1+m^2}$$

is as large as possible. To find the value of m graphically, set the range to be [0, 3] x [0, 1000] with the x scale equal to 1 and the y scale equal to 100 and draw the graph of

Y=(968X+44√(487X²+3))/(1+X²).

The graph will appear as in the figure at the top of the next page.

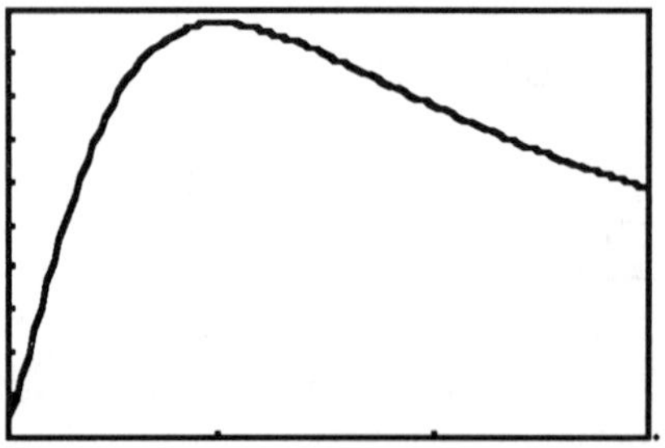

Zooming in we find the maximum point of this graph occurs at about x = 0.9969. So shooting the projectile with a slope of m = 0.9969 gives the maximum horizontal distance.

To see this graphically, execute the program given at the beginning of this example with the range setting [0, 1000] x [0, 400], both x and y scales set equal to 100. Taking **V** = 176, **H** = 3 and **M** = 0.5, 0.75, 0.9969, 1.5, and 2 successively, the graphs will look something like this:

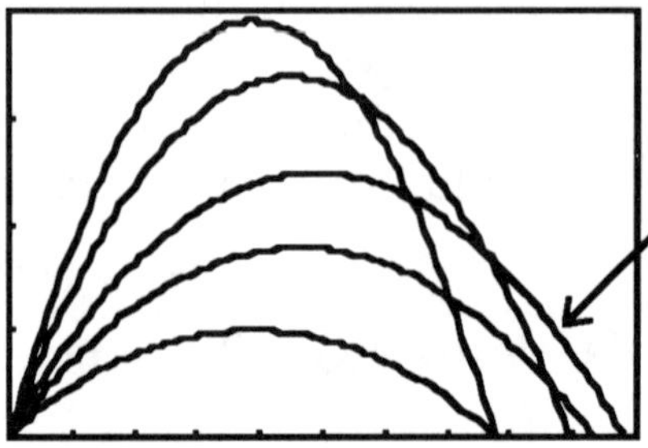

5.1 Exercises

For exercises 1 - 9 solve each equation using the quadratic formula.

1. $2x^2 + 9x + 9 = 0$
2. $3x^2 + 11x - 20 = 0$
3. $10x^2 + 33x - 7 = 0$
4. $x^2 - 4x + 1 = 0$
5. $x^2 - x - 1 = 0$.
6. $9x^2 - 6x - 17 = 0$
7. $x^2 - 6x + 13=0$
8. $25x^2 - 20x + 7 = 0$
9. $x^2 - \sqrt{2}x + 1 = 0$

For exercises 10 - 12 solve each equation using an appropriate change of variable and th solving the resulting quadratic equation.

10. $3x^4 + 16x^2 - 35 = 0$
11. $x - 4\sqrt{x} - 3 = 0$
12. $3 \cdot 9^x - 82 \cdot 3^x + 27 = 0$

13. In Chapter 1, section 4 the formulas for the volume and surface area of right circular cylinder of radius **R** and height **H** were given. Recall

    Volume: $\mathbf{V} = \pi \mathbf{R}^2\mathbf{H}$

Surface Area: $\mathbf{S = 2\pi R^2 + 2\pi RH}$

If the values of H, V, and S are fixed, the above formulas become quadratic equations in the variable R:

$$\pi HR^2 - V = 0$$

$$2\pi R^2 + 2\pi HR - S = 0.$$

(a) Fix the value of **H** = 5. Find **R** so that **V** = 1000. Compare your answer with the value obtained in exercise 2, section 1.4.

(b) Fix the value of **H** to be 5. Find **R** so that **S** = 100.

(c) Solve for **R** algebraically with the quadratic formula, thereby giving the formulas for **R** when **H**, **V**, and **S** are fixed values. Draw graphs of these relations.

14. Suppose a baseball is thrown straight up from a height of 4.5 feet with an initial velocity of 97 feet per second. Find when the baseball hits the ground and how high it goes.

15. A projectile is shot from a height h_0 with an initial velocity v_0 and an upward slope of m, but now there is a wind of velocity w. The path of the projectile is given by the quadratic equation

$$y = \frac{-16\left(1 + m^2\right)}{\left(v_0 - w\sqrt{1 + m^2}\right)^2} x^2 + \frac{v_0 m}{v_0 - w\sqrt{1 + m^2}} x + h_0.$$

If $w > 0$ then the projectile is shot into the wind and if $w < 0$ then the projectile is shot with the wind. Change the programs given in Example 5.1.4 to prompt for the input of w, compute the values of $\mathbf{A} = -16(1+\mathbf{M}^2)/(\mathbf{V}-\mathbf{W}\sqrt{1 + \mathbf{M}^2})^2$, $\mathbf{B} = \mathbf{VM}/(\mathbf{V}-\mathbf{W}\sqrt{1 + \mathbf{M}^2})$, and **C** = **H**, and draw the graph of the projectile's path.

(a) For **H** = 3, **V** = 176 and **W** = 15 determine **M** so that the projectile travels the farthest.

(b) Determine the horizontal distance travelled by a projectile shot with a slope of 1, and initial velocity of 97, a height of 5, but with no wind.

(c) Suppose in part (b) there is a 25 feet per second wind in the same direction as the projectile, i.e., **W** = -25. Determine **M** so that the projectile travels the same distance as in part (b).

(d) Here is a game. Fix the height and the wind velocity and its direction as well. In three attempts find an initial velocity and slope so that the projectile travels approximately 1000 feet. For example, suppose **H** = 5 and **W** = 22. Taking **M** = 1.25 and **V** = 200 gives the distance travelled as 1008.06 feet. The person closest after three guesses wins.

5.2 Quadratic Inequalities

(McKeague: Section 7.5)

A quadratic inequality is an exprssion of the type

$$x - 8 > x^2 - 4.$$

The goal here is to develop a graphical meaning to this expression and determine its solution sets.

Most inequalities can be viewed as one of the following types: $f(x) \geq g(x)$ or $h(x) \geq 0$. (Of course the symbols > or < would also be used if appropriate.). Graphically what the solution sets mean: all values of x where the graph of f(x) is above the graph of g(x) or the graph of h(x) is above the x-axis. Each of the three inequalities given above is in one of the basic forms and the solution sets can be interpreted graphically as stated. Let's illustrate this with some examples.

- **Example 1:** Solve the inequality $x^2 - 2x - 8 < 0$. (McKeague: Example 1, Section 7.5)

By factoring the quadratic equation $x^2 - 2x - 8 = 0$ we find $x = 4$ or $x = -2$. So the solution set of the inequality is found by testing the intervals: $(-\infty, -2)$, $(-2, 4)$, and $(4, +\infty)$. We find that the solution set is $-2 \leq x \leq 4$. To determine the solution graphically; draw the graph of $y = x^2 - 2x - 8$. Using the range $[-5, 5] \times [-15, 15]$ gives the graph

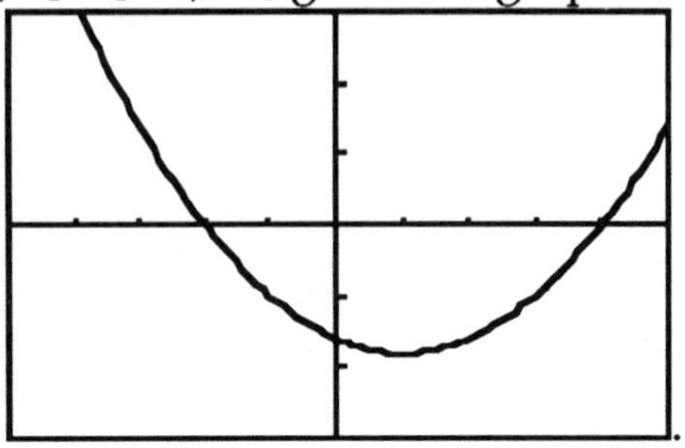

We see from the graph that $y = x^2 - 2x - 8$ is below the x-axis if $-2 < x < 4$. Why are -2 and 4 included in the solution set?

- **Example 2:** A football is kicked straight up from a level field from a height of 3.5 feet and with an initial velocity of 80 feet per second. During what time interval will the football be at least 80 feet above the ground?

The governing equation giving the height above the ground at any time t is

$$h = -16t^2 + 80t + 3.5.$$

We wish to solve the inequality $h \geq 80$, i.e., $-16t^2 + 80t + 3.5 \geq 80$. Drawing the graphs $y = -16x^2 + 80x + 3.5$ and $y = 80$ with the range $[0, 6] \times [0, 110]$ produces

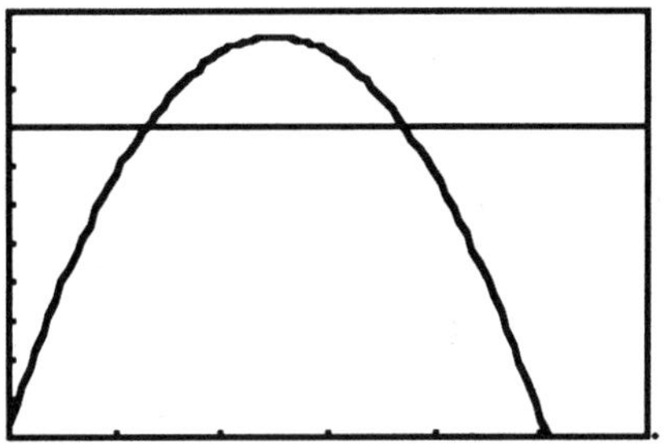

Zooming in a few times on the points of intersection produces the graphical solution $1.28 \le t \le 3.71$. Using the quadratic formula program on the equation $-16t^2 + 80t - 76.5 = 0$ gives the solutions $1.288080036 \le t \le 3.711919964$. This equation is found by setting $-16t^2 + 80t + 3.5$ equal to 80.

5.2 Exercises

Solve the following inequalities graphically and algebraically.

1. $3x^2 - 4x - 15 < 0$.
2. $10 + 3x - x^2 \ge 0$.
3. $|5x^2 - 8x - 2| \ge 5$.
4. Determine the values of the radius R so that the surface area of a right circular cylinder of height 5 feet is at least 150 square feet.
5. Suppose a baseball is thrown vertically upward from level ground from a height of 5 feet and with an initial velocity of 125 feet per second. Find the time interval where the ball is at least 150 feet above the ground.
6. In exercise 15 of section 5.1 the quadratic equation for the path of a projectile with an influencing wind is given. Suppose a projectile is shot from a height of 15 feet with an initial velocity of 187 feet per second and into a wind of 22 feet per second. Find graphically the values of the slope $m > 0$ so that the projectile travels a horizontal distance of at least 500 feet but not more than 550 feet. (Hint: Start with $m = 1$ to see what the approximate maximum distance is and then go from there. The slopes could be greater than or less than 1.)

5.3 Graphing Parabolas

(McKeague: Sections 7.6, 7.5)

Any function of the form

$$f(x) = ax^2 + bx + c$$

for real constants $a \ne 0$, b, and c is called a quadratic function. The graph of such a function is given a special name, a parabola. As we have seen before, all quadratic functions have graphs that look like one of two basic shapes. These are illustrated in the graph at the top of the next page.

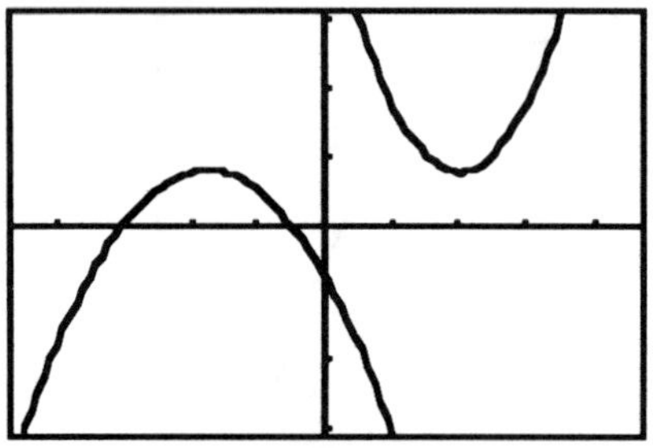

They either open downward or open upward. Of course some may have x-intercepts and others may not. The lowest or highest point of the parabola is called the vertex. Before we found this point using trace and zooming in techniques, but this, as you know already, can take quite some time. With a little algebra it is possible to find the vertex in terms of the values of a, b, and c.

By completing the square it can be shown that

$$f(x) = ax^2 + bx + c = a\left(x - \left(-\frac{b}{2a}\right)\right)^2 + \left(c - \frac{b^2}{4a}\right).$$

So the vertex has the coordinates $\left(-\frac{b}{2a}, c - \frac{b^2}{4a}\right)$.

- **Example 1:** Determine how high a projectile will travel if shot with an upward slope of 1.5, from a height of 4.5 feet, and with an initial velocity of 221 feet per second.

From example 14, section 5.1, we know that the path of the projectile is given by $y = ax^2 + bx + c$ where

$$a = \frac{-16\left(1 + (1.5)^2\right)}{221^2} = -0.0010646793, \ b = 1.5, \text{ and } c = 4.5.$$

So the vertex or highest point of the path is x = 704.4375, y = 532.828125. Therefore the projectile goes approximately 532.83 feet high. The graph of the path is

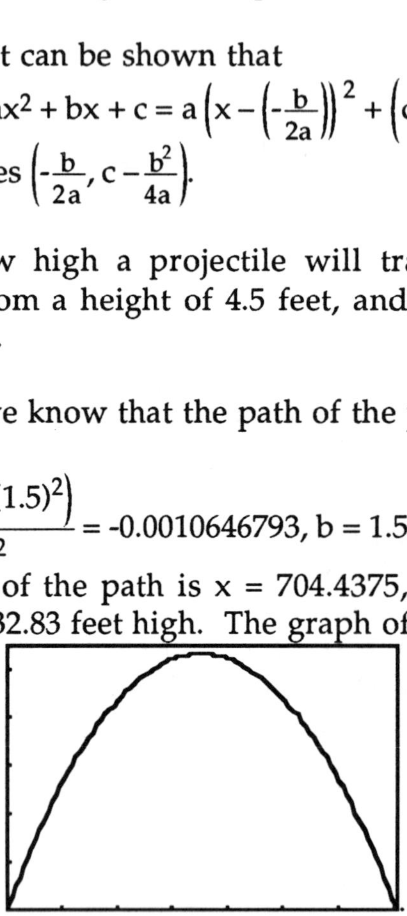

We know that if $a > 0$ the parabola opens upward and if $a < 0$ the parabola opens downward. Using this fact along with knowledge of the coordinates of the vertex it is possible to set the range so that the parabola can be viewed fairly well. The following programs accomplish this task. These programs will also draw parabolas that open left and right.

CASIO

Program 5.3.1C: Parabola grapher

Keystrokes	Explanation
"TYPE?" ↵	Prompts for the input of the type of parabola.
"1–VERT" ↵	Enter 1 if the parabola opens up or down. Enter 2 if
"2–HORIZ":?→T: ↵	the parabols opens left or right.
"A"?→A: ↵	Prompts for the input of the coefficients a, b, and c.
"B"?→B:"C"?→C ↵	
T=2⇒Goto 1 ↵	
-B÷2A→H ↵	Here the parabola opens up or down. Compute the
C–B²÷4A→K ↵	x coordinate of the vertex and store it in **H** and
Goto 2 ↵	compute the y coordinate of the vertex and store it in **K**.
Lbl 1:C–B²÷4A→H ↵	Here the parabola opens left or right. Compute the
-B÷2A→H ↵	x coordinate of the vertex and store it in **H** and compute the y coordinate of the vertex and store it in **K**.
Lbl 2: "H=":H ◢	Display the coordinates of the vertex.
"K=":K ◢	
Prog_ ↵	**Prog_** should be the program location of the range setting program, 2.1.1C.
T=2⇒Goto 3 ↵	If the parabola opens left or right we go to label 3 and draw the appropriate graph.
Graph Y=AX²+BX+C ↵	Graph the parabola the opens up or down.
Goto 4 ↵	Go to label 4 and end the program.
Lbl 3 ↵	Here the parabola opens left or right. To graph
Graph Y=√((X–H)÷A)+K ↵	$x = ay^2 + by + c$, we complete the square and solve
Graph Y=-√((X–H)÷A)+K ↵	for y. Giving the expressions shown here.
Lbl 4	

TI-81/82
The equivalent TI-82 commands are in parentheses

Program 5.3.1T: PARABG, Parabola Grapher

Commands	Explanation
:All–Off (FnOff)	
:Disp "TYPE?"	Prompts input for the type of parabola. Enter 1 if
:Disp "1–VERT"	the parabola opens up or down and enter 2 if the
:Disp "2–HORIZ"	parabola opens left or right.
:Input T	
:Disp "A"	Prompts for the input of the coefficients a, b, and c.

`:Input A`
`:Disp "B"`
`:Input B`
`:Disp "C"`
`:Input C`	

`:If T=2` `:Goto 1`	Here the parabola opens left or right so go to label 1.
`:-B/2A→H`	Here the parabola opens up or down. Compute the x coordinate of the vertex and store it **H**.
`:C–B2/4A→K` `:Goto 2`	Compute the y coordinate of the vertex and store it in **K**.
`:Lbl 1`	Here the parabola opens left or right.
`:C–B2/4A→H`	Compute the x coordinate of the vertex and store it in **H**.
`:-B/2A→K`	Compute the y coordinate of the vertex and store it in **K**.
`:Lbl 2`	

`:Disp "H"`
`:Disp H`
`:Disp "K"`
`:Disp K`
`:Pause`
	Displays the values of **H** and **K**.

`:Prog_`	**Prog_** should be the program location of the range setting program, 2.1.1T.
`:If T=2` `:Goto 3`	If the parabola opens left or right we go to label 3 and draw the appropriate graph.
`:"AX2+BX+C"→Y1` `:Y1–On (FnOn 1)`	Set the expression for the up or down parabola in **Y1**.
`:Goto 4`	Go to label 4 and display the graph.

`Lbl 3 ↵`
`"√((X–H)/A)+K"→Y1`
`"-√((X–H)÷A)+K"→Y2`
`:Y1–On (FnOn 1,2)`
`:Y2–On`
`:Lbl 4`
`:DispGraph`
	Here the parabola opens left or right. To graph $x = ay^2 + by + c$, we complete the square and solve for y. Giving the expressions shown here.

• **Example 2:** A company selling copies of an accounting program for home computers finds that it will make a weekly profit of P dollars from selling x copies of the program according to the equation

$$P = -0.1x^2 + 27x - 500$$

How many copies of the program should it sell to make the largest possible profit, and what is the largest possible profit.

(McKeague: Example 5, Section 7.6)

Executing the appropriate program given above, enter 1 for the type of parabola and **A** = -0.1, **B** = 27, and **C** = -500 gives for the vertex **H** = 135, **K** = 13.22.5. When the range program is executed enter **H** for **X** and **K** for **Y** and 1 for **S**. The graph produced by the program is

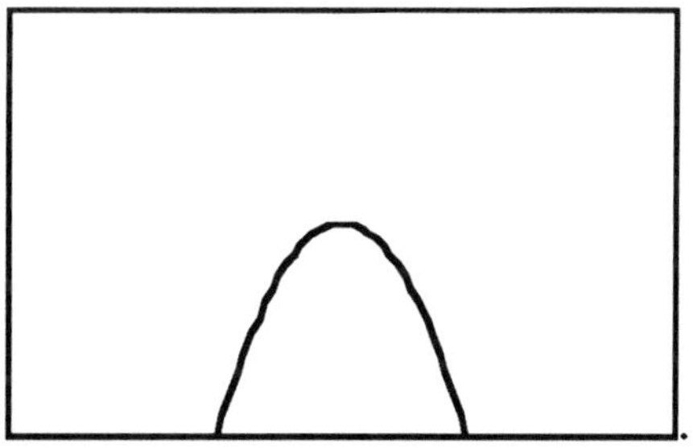

So the maximum profit is $1322.50 if they sell 135 copies. Certainly the company can choose the sell any number of copies but in order for them to make a profit, they should choose x so that P > 0. Using the quadratic formula we find the P > 0 if 20 < x < 250. Setting the range to be [0, 250] x [0, 1323] and drawing the graph will show the entire graph of P that is of interest. This graph is shown below.

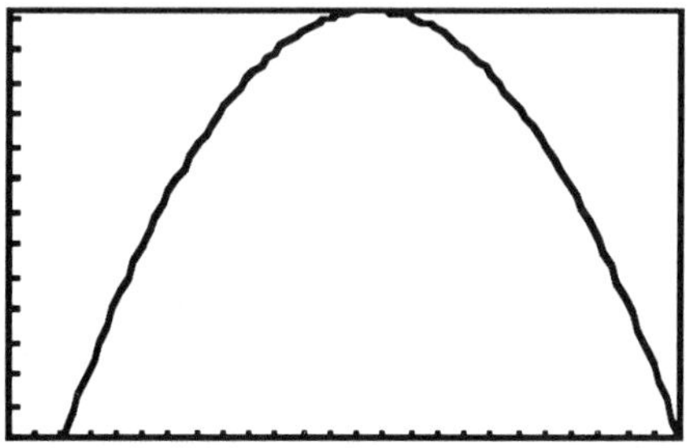

• **Example 3:** Determine the vertex of the parabola whose equation is $x = \frac{1}{4}y^2 - 3y + 2$. Draw the graph of the parabola.

Executing the appropriate program given above enter 1 for the type of parabola and **A** = 0.25, **B** = -3, and **C** = 2 gives for the vertex **H** = -7, **K** = 6. When the range program executes enter -7 for X, 6 for Y, and 0.2 for S. The graph produced is given at the top of the next page.

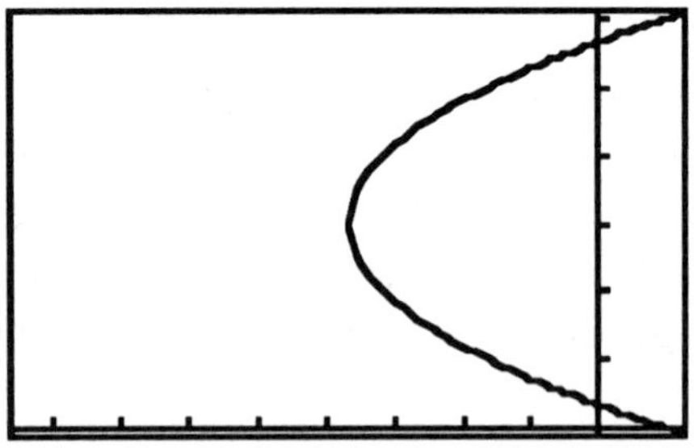

From a graphical point of view all parabolas are simple rescalings and translations of the simple parabola $g(x) = x^2$. To see this recall that

$$f(x) = ax^2 + bx + c$$
$$= a(x - h)^2 + k$$

where $h = -b/2a$ and $k = c - b^2/4a$. Thus to produce the graph of $f(x)= ax^2 + bx + c$, take the graph of $g(x) = x^2$ and translate to the left or right by an amount $|h|$, rescale it by a, and move it up or down by an amount $|k|$.

• **Example 4:** By rescaling and translating the graph of $g(x) = x^2$ construct the graph $f(x) = -3x^2 + 6x - 1$.

By completing the square or by the programs given above we find the vertex to be (1, 2). So we rescale $g(x) = x^2$ by -3, translate it to the right 1 unit, and move it up by 2 units. This process is depicted in the four figures below.

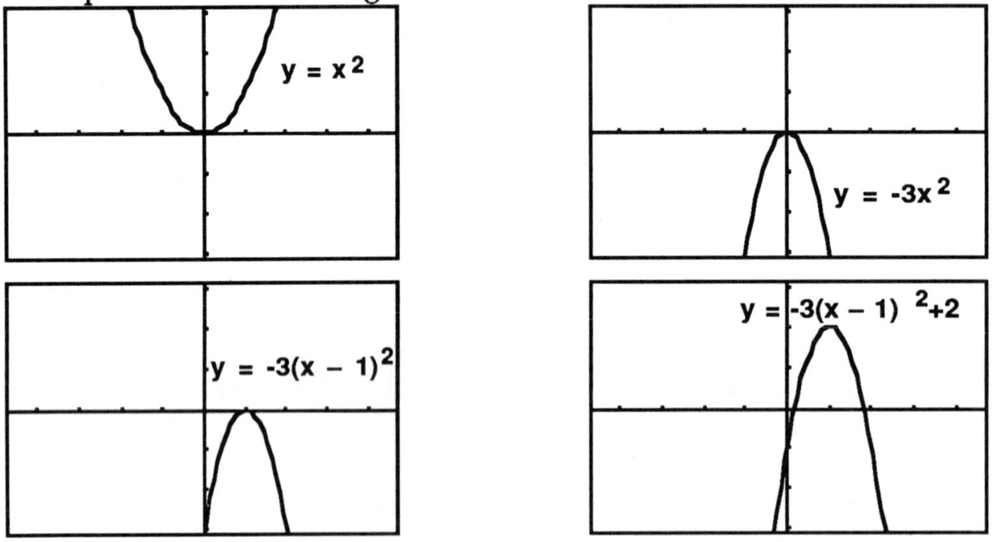

5.3 Exercises

Find the vertex of each parabola and draw its graph. Determine how the graph of each can be found by rescaling and translating the graph of $y = x^2$.

1. $f(x) = 2x^2 + 8x + 5.$
2. $f(x) = -x^2 + 5x - 4.$
3. $f(x) = \frac{1}{4}x^2 + 2x + 1.$
4. $f(x) = -\frac{2}{3}x^2 - 8x + 8.$

Determine the vertex of each parabola and draw its graph.

5. $y = 2x^2 - 4x + 1.$
6. $y = -x^2 - 2x + 1.$
7. $y = -5x^2 + 15x - 11.$
8. $x = \frac{1}{3}y^2 - y.$
9. $y = \frac{1}{4}x^2 + x - 3.$
10. $x = -y^2 + y - 5.$
11. $x = 4y^2 - 4y + 10.$
12. $x = -\frac{1}{10}y^2 + y - 1.$

13. Refer to exercise 15, section 5.1. A projectile is shot at a slope of 1.25, from a height of 6 feet with an initial velocity of 121 feet per second.
    (a) Find the maximum height travelled by the projectile.
    (b) Suppose it is shot into a headwind with a velocity of 15 feet per second. How does this affect the maximum height?
    (c) Suppose instead it is shot with a tailwind with a velocity of 27 feet per second. How does this affect the maximum height?
    (d) Is there any difference in the maximum height for the different wind velocities and wind directions? Prove your answer.
    (e) What happens when the wind velocity increases to about 60% of the initial velocity? Can you find what wind velocity will force the projectile to go nearly straight up and straight down? Ans: When the wind velocity is approximately 75.588 feet per second the horizontal distance travelled is approximately 0.0006 feet.

14. From the graphs of quadratic functions $f(x) = ax^2 + bx + c$ it is easy to see that the value of a determines which way the parabola opens and also how wide or narrow it opens. The value of c simply translates the parabola up or down, but what does the value of b do? Fix $a = 1$ and $c = 1$ and draw several parabolas for different values of b between -3 and 3. What changes occur with each value of b? Now repeat the process, but only plot the vertex for each value of b. What is the curve that the vertices create? Can you find the equation of this curve?

Investigating Quadratic Inequalities

Think about how you could use the graphing capabilities of the graphing calculator to solve inequalities.

Solve for x: $x^2 - 2x \leq 8$.

Set each side of the inequality equal to y and graph on the same axis as shown in Figure 1.

$$y_1 = x^2 - 2x$$

$$y_2 = 8$$

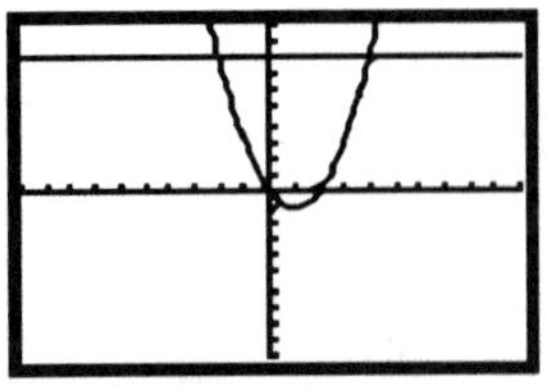

Figure 1

The original inequality states that $y_1 \leq y_2$ so you are interested in the x-values where that is true. Using TRACE and analyzing the graph, you find the answer is

$$-2 \leq x \leq 4.$$

Write an algorithm for graphing inequalities using graphical techniques. Use your algorithm to solve these three problems.

1. $x^2 - 3x - 10 \leq 0$
2. $2x^2 + 5x > 7$
3. $x^3 - 6 \leq 7x$

Chapter 6

Systems of Linear Equations

A great deal of the computations done everyday involve solving a system of more than one linear equation. Very large systems are used to determine the best strategies for allocating resources in the airline and communication industries and much of the time used on supercomputers is devoted to solving large systems of equations. In this chapter we will learn how the visualization of linear systems of equations and inequalities aids in the solution of such systems.

6.1 Systems of Two Linear Equations (Mckeague: Section 8.1)

In this section we explore how using the zoom feature on the calculators and the range settings can help in determining approximate solutions to systems of two linear equations.

Any solution of the system

$$2x - y = 1$$
$$3x + 5y = 5$$

must be a point where the graphs of the two equations intersect. Solving both for y gives the functions

$$y = 2x - 1 \text{ and } y = -\frac{3}{5}x + 1.$$

Graphing these functions using the default range gives

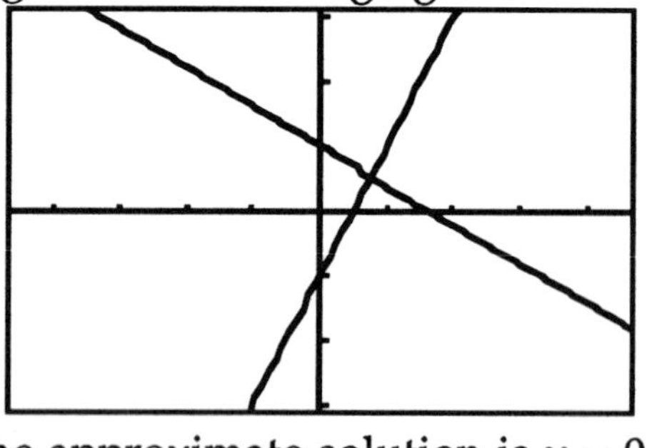

Using the trace, we find that the approximate solution is x = 0.8, y = 0.52. Substituting these values into the left side of the equations gives

$$2(0.8) - 0.52 = 1.08 \text{ and } 3(0.8) + 5(0.52) = 5.$$

Note that (0.8, 0.52) lies on the second line, but not on the first. Also the point nearly satisfies the first equation, indicating that it does approximate the true solution. If we zoom in on this point by a factor of 10 we find x = 0.77 and y = 0.54. Trying these values in the equations gives

$$2(0.77) - 0.54 = 1 \text{ and } 3(0.77) + 5(0.54)=5.01.$$

If we zoom in twice more by a factor of 10 then x = 0.769, y =0.538. Checking the range we find

CASIO: [0.7643, 0.7737] x [0.5359, 0.5421]

TI-81:

[0.76475, 0.77425] x [0.53535, 0.54165].

We see here the **Xmax** and **Xmin** and **Ymax** and **Ymin** differ by at most 0.009. This tells us that the pair (0.769, 0.538) approximates the true solution to about one decimal place.

The differences Xmax – Xmin and Ymax – Ymin give the error in the approximation. For example, if both of these differences are less than 0.01, then the error in the approximate solutions is at most 0.01.

Substituting these values into the equations gives

$$2(0.769) - 0.538 = 1 \text{ and } 3(0.769) + 5(0.538) = 4.997.$$

This is fairly good agreement.

The exact values are found by using the elimination method. From the first equation we have that y = 2x – 1. Substituting this expression for y in the second equation gives

$$3x + 5(2x - 1) = 5$$
$$13x - 5 = 5$$
$$13x = 10$$
$$x = \frac{10}{13} = 0.7692307692.$$

So y = 2(10/13) – 1 = 7/13 = 0.5384615385.

6.1 Exercises

Find approximate solutions of each system by zooming in on the intersection points of the graphs until the error is 0.001. Substitute these approximate solutions into the equations to see how well they satisfy the system. Using the elimination method find the exact solutions.

1. $-x + 4y = 1$
   $2x + y = 3.$
2. $\frac{1}{2}x - 3y = -8$
   $7x + 2y = 10.$
3. $11x + 10y = 2$
   $12x + 9y = 1.$
4. $2x - 3y = 1$
   $3x + 2y = 5.$

Solving Systems of Linear Equations Using Matrices

Mathematics is the study of patterns. Often we can use a known pattern to help us solve a new problem. Consider the pattern for solving a linear equation of the form

$$ax = b$$

where x is the unknown. To solve for x multiply both sides of the equation by the multiplicative inverse of a, a^{-1}.

$$(a^{-1})\,ax = (a^{-1})\,b$$

$$x = \frac{b}{a}$$

Consider the system of linear equations

$$\begin{aligned} 2x - 3y &= 13 \\ 4x + 5y &= -7 \end{aligned}$$

which can be written as the matrix equation

$$\begin{pmatrix} 2 & -3 \\ 4 & 5 \end{pmatrix} \bullet \begin{pmatrix} x \\ y \end{pmatrix} = \begin{pmatrix} 13 \\ -7 \end{pmatrix}$$

Observe that this equation follows the pattern AX = B. So, $X = A^{-1} \cdot B$.

To find X begin by entering the matrix $A = \begin{pmatrix} 2 & -3 \\ 4 & 5 \end{pmatrix}$ into your calculator.

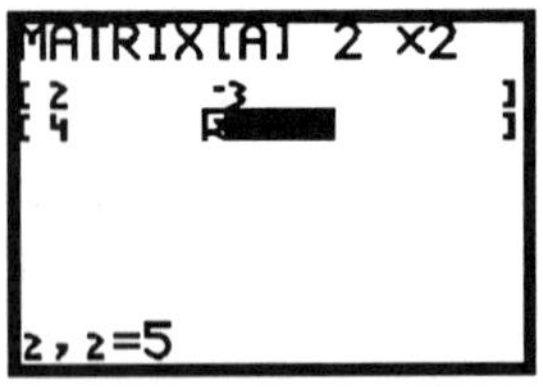

Then enter the matrix $B = \begin{pmatrix} 4 \\ 3 \end{pmatrix}$ into your calculator.

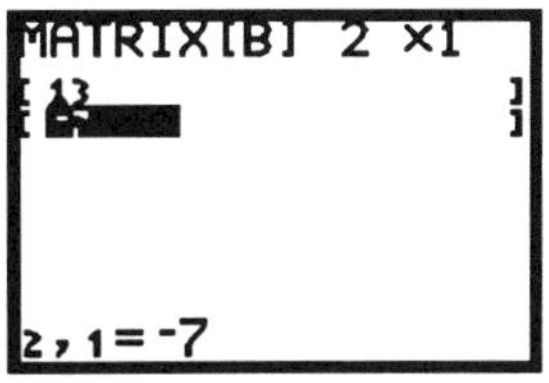

To solve for X, enter $A^{-1}*B$ on the normal screen of the calculator.

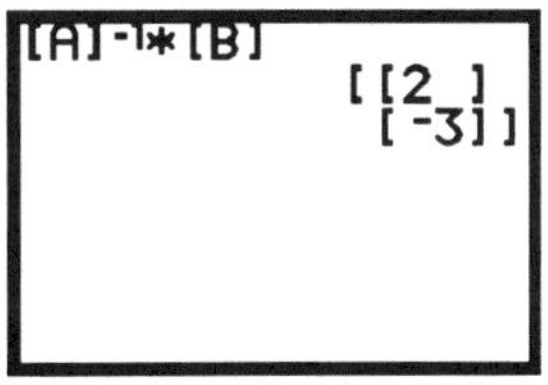

The answer $\begin{pmatrix} 2 \\ -3 \end{pmatrix}$ is interpreted as $\begin{pmatrix} x \\ y \end{pmatrix} = \begin{pmatrix} 2 \\ -3 \end{pmatrix}$ where x = 2 and y = -3.

This method can be used for the following 3x3 linear system.

$$\begin{aligned} x + y - 4z &= -1 \\ 3x - 2y + 6z &= -9 \\ 2x + 3y - 2z &= 4 \end{aligned}$$

The matrix equation to solve this system would be

$$\begin{pmatrix} 1 & 1 & -4 \\ 2 & 3 & -2 \\ 3 & -2 & 6 \end{pmatrix} \bullet \begin{pmatrix} x \\ y \\ z \end{pmatrix} = \begin{pmatrix} -1 \\ 4 \\ -9 \end{pmatrix}$$

Observe this equation also follows the pattern AX = B, where $X = A^{-1}*B$.

The result is $\begin{pmatrix} x \\ y \\ z \end{pmatrix} = \begin{pmatrix} -2 \\ 3 \\ .5 \end{pmatrix}$, which means that x = -2, y = 3, and z = 0.5.

The CASIO fx-7700G can be used to solve systems of n equations in n unknowns using the inverse of the coefficient matrix. This is accomplished using the **MATRIX** calculator mode (**MODE** **0**). To enter the coefficient matrix

$$A = \begin{pmatrix} 1 & 1 & -4 \\ 2 & 3 & -2 \\ 3 & -2 & 6 \end{pmatrix}$$

press **F1** (A). If the dimensions are not 3x3 or it doesn't exist then press **F6** (▽) followed by **F1** (**DIM**). Enter 3 for both **Row:** and **Colm:** pressing **EXE** after each. Enter each entry of the first row of the matrix followed by **EXE**. You now have the coefficient matrix entered. Press **PRE** and follow the same procedure to enter the right hand side of the system

$$B = \begin{pmatrix} -1 \\ 4 \\ -9 \end{pmatrix}.$$

When completed press **PRE** followed by **F1** (A). Press **F4** (A^{-1}). This will compute the inverse of the coefficient matrix and put it into matrix C. Press **F1** (C→A). This will put the inverse of A back into A. Press **PRE** followed by **F5** (X). This will multiply A (which is the inverse of the coefficient matrix) and B (the right hand side) and put the product into C (the solution). The solution is now given in C.

$$C = \begin{pmatrix} -2 \\ 3 \\ .5 \end{pmatrix}.$$

✎ *Research the definition of multiplication of matrices and explain how the given matrix equations do produce the given systems of equations.*

✎ *Use the method of matrix equations to solve the following three systems. Discuss when you would use this method instead of the traditional paper-pencil method.*

1. $2x - 3y = 3$
   $4x - 2y = 10$

2. $-x + y + 3z = 6$
   $x + y + 2z = 7$
   $2x + 3y + z = 4$

3. $2x - 3y + z = 1$
   $3x - y - z = 4$
   $4x - 6y + 2z = 3$

Chapter 7

Functions and Conic Sections

The development of visualizing algebraic objects in a coordinate system is one of the most useful creations of mathematics. In this chapter we will look at some basic ideas of what has become analytic geometry and how the graphics calculator can facilitate its study.

7.1 Explorations

(McKeague: Sections 9.1, 9.2)

Let f(x) be some function given by an algebraic formula, for example $f(x) = \sqrt{x}$. We will explore here the effect on the graph of f(x) when the numbers a, b, c, and d are varied in the expression af(b(x+c))+d.

- Example 1: What is the effect on the graph of $f(x) = \sqrt{x}$ when the values of a, b, c, and d are varied in the function $g(x) = a\sqrt{b(x + c)} + d$?

CASIO
Create the following simple program.

```
"A"?→A:"B"?→B ↵
"C"?→C:"D"?→D ↵
Graph Y = A√(B(X+C))+D ↵
```

TI-81/82
Create the following simple program.

```
:Disp "A"
:Input A
:Disp "B"
:Input B
:Disp "C"
:Input C
:Disp "D"
:Input D
:DrawF A√(B(X+C))+D
```

Store the values **A** = 1, **B** = 1, **C** = 0, and **D** = 0. With the default range setting draw the graph of g(x), which for these values of **A, B, C,** and **D** will be $\sqrt{x}$.

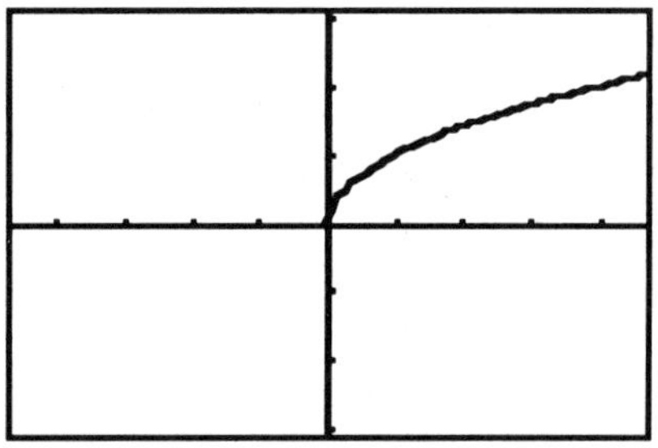

Since $\sqrt{x}$ exists only for $x \geq 0$ and $\sqrt{x} \geq 0$ the graph appears only in the first quadrant. Without clearing the graphics screen and keeping the previous values of **A**, **B**, and **C** the same, draw the graphs when **D** = -2, $\frac{1}{2}$, and 2. We get

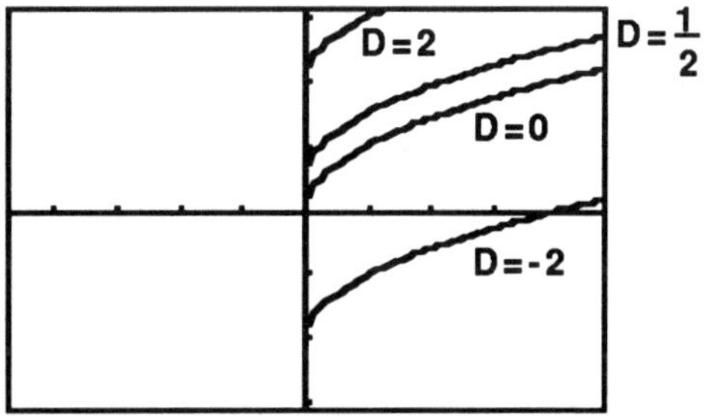

We see that changing **D** moves the graph of $f(x) = \sqrt{x}$ up or down. Now set **D** = 0, **C** = 0, **B** = 1 and vary **A**. For **A** = 1, 2, -1, and $\frac{1}{2}$ we get the graphs

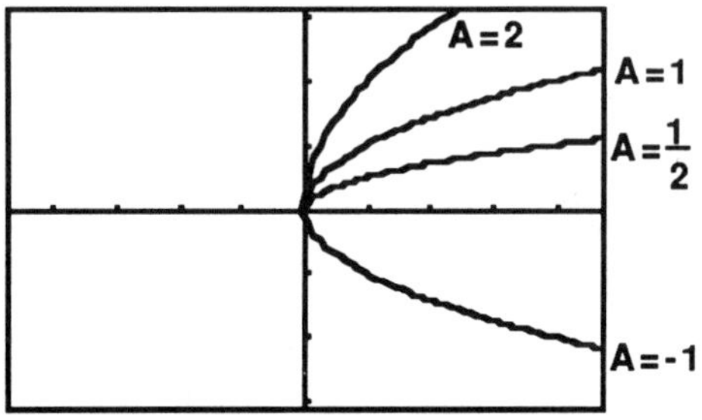

We see that **A** changes how the function increases, and if **A** < 0 it can also reflect it across the x-axis. To see this in a different light, draw the graph of $f(x) = \sqrt{x}$ on the default range setting and study it for a moment. Now change the range to

CASIO: [-4.7, 4.7] x [-6.2, 6.2], **Xscl** = **Yscl** = 1
TI-81: [-4.8, 4.7] x [-6.4, 6.2], **Xscl** = **Yscl** = 1
TI-82: [-4.7, 4.7] x [-6.2, 6.2], **Xscl** = **Yscl** = 1

and draw $g(x) = 2\sqrt{x}$. What do you notice? Does it appear to look the same as $f(x) = \sqrt{x}$ on the default range setting? The way to describe the effect of **A** is that it rescales the y-axis.

Reset the range now to the default setting and set **A** = 1, **B** = 1, **D** = 0, and vary **C**. For **C** = 0, -3, 1, $-\frac{1}{2}$, and 2 we get the graphs at the top of the next page.

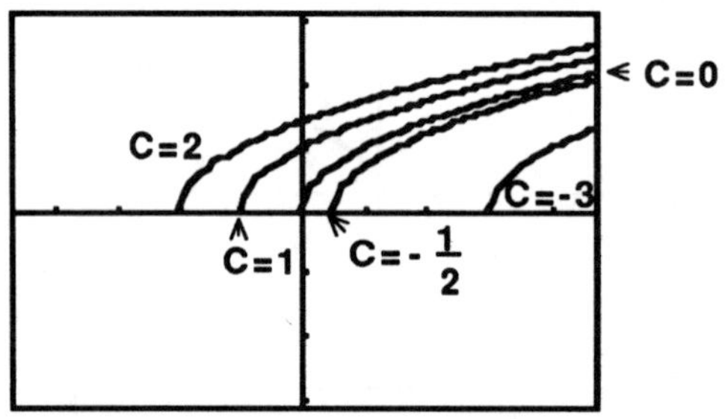

We see that the graph is translated to the left if **C** > 0 and translated to the right if **C** < 0.

Suppose we set **A** = 1, **C** = 0, **D** = 0 and vary **B**. What do you suppose will happen? Draw f(x) = $\sqrt{x}$ with default setting and remember its looks and then draw g(x) = $\sqrt{\frac{1}{2}x}$ with the range settings

CASIO: [-9.4, 9.4] x [-3.1, 3.1], **Xscl** = **Yscl** = 1,
TI-81: [-9.6, 9.4] x [-3.2, 3.1], **Xscl** = **Yscl** = 1,
TI-82: [-9.4, 9.4] x [-3.1, 3.1], **Xscl** = **Yscl** = 1.

What do you notice? Changing the value of **B** will rescale the x-axis. What will happen if **B** < 0?

To review consider the function g(x) = $2\sqrt{-3(x+2)} - 2.5$. This is the function f(x) = $\sqrt{x}$ shifted down 2.5 units, reflected across the y-axis, translated to the left 2 units, the x-axis rescaled by $\frac{1}{3}$ and the y-axis rescaled by 2. The figure below depicts both graphs drawn on the screen using the range settings

CASIO: [-9.4, 9.4] x [-6.2, 6.2], **Xscl** = **Yscl** = 1,
TI-81: [-9.6, 9.4] x [-6.4, 6.2], **Xscl** = **Yscl** = 1,
TI-82: [-9.4, 9.4] x [-6.2, 6.2], **Xscl** = **Yscl** = 1

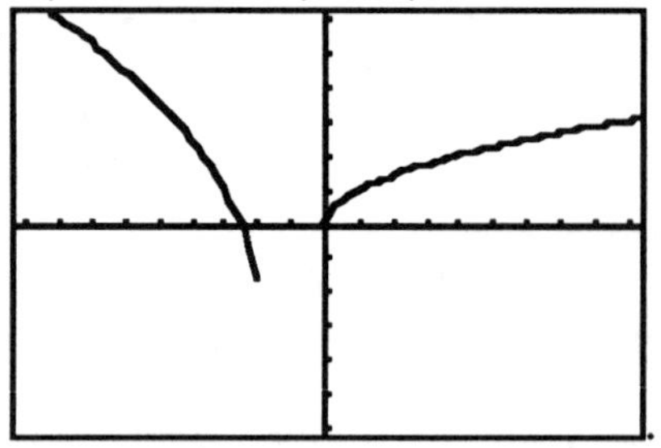

The other method we can use to create new functions from old ones is to apply the arithmetic operations to two or more functions. That is, given f(x) and g(x) we can create the new functions (f ± g)(x), (fg)(x), (f/g)(x). The other operation that is often used is composition, $(f \circ g)(x) = f(g(x))$.

- **Example 2:** Given f(x) = 2x – 1and g(x) = 4 – x^2, draw the graphs of (f ± g)(x), (fg)(x), (f/g)(x), (g/f)(x), f[g(x)], and g[f(x)].

CASIO

Set the range to be [-9.4,9.4] x [-6.2,6.2]. In succession draw the graphs of $y_1 = 2x - 1$, $y_2 = 4 - x^2$, $y_3 = 2x - 1 + 4 - x^2$, and $y_3 = 2x - 1 - 4 + x^2$. The screen should appear as,

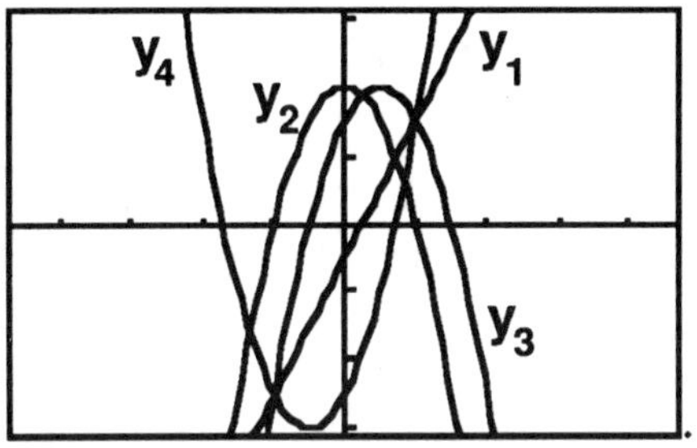

Reset the range to be [-5, 5] x [-10, 10] and draw in succession the graphs $y_1 = 2x - 1$, $y_2 = 4 - x^2$ and $y_5 = (2x - 1)(4 - x^2)$. The screen should appear as

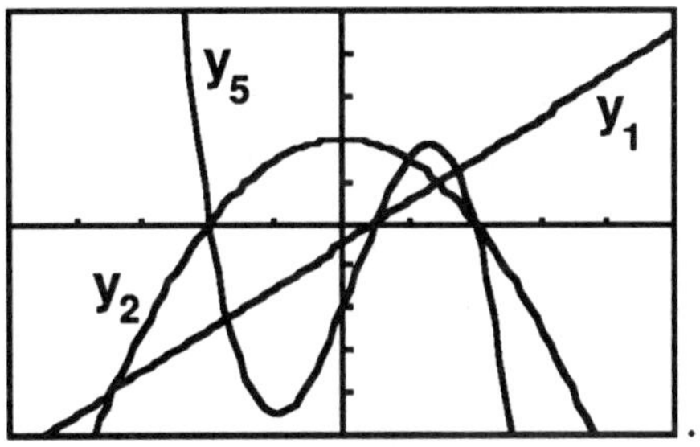

Repeating a similar procedure for $y_1 = 2x - 1$, $y_2 = 4 - x^2$, $y_6 = y_1/y_2$, and $y_7 = y_2/y_1$, but using the range [-5, 5] x [-8, 8] we get

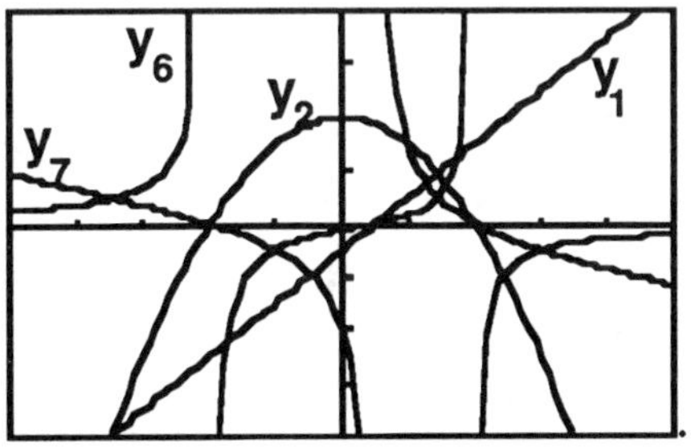

Why does the graph of y_6 run off the screen at $x = \pm 2$? Checking the value of y_7 at $x = 1.99$ we find $y_6 = 74.68671679$ and at $x = 2.01$, $y_7 = -75.3117207$. We see that for x near 2, the value of $|y_6|$ increases without bound. Why does the graph of y_7 run of the screen at $x = 1/2$?

For the compositions we have, if $f(x) = y_1 = 2x - 1$ and $g(x) = y_2 = 4 - x^2$, the functions

$$y_8 = f[g(x)] = 2(4 - x^2) - 1 = -2x^2 + 7$$

$$y_9 = g[f(x)] = 4 - (2x - 1)^2 = 4 - (4x^2 - 4x + 1) = -4x^2 + 4x + 3.$$

Drawing these graphs with the range [-5, 5] x [-8, 8] you should see

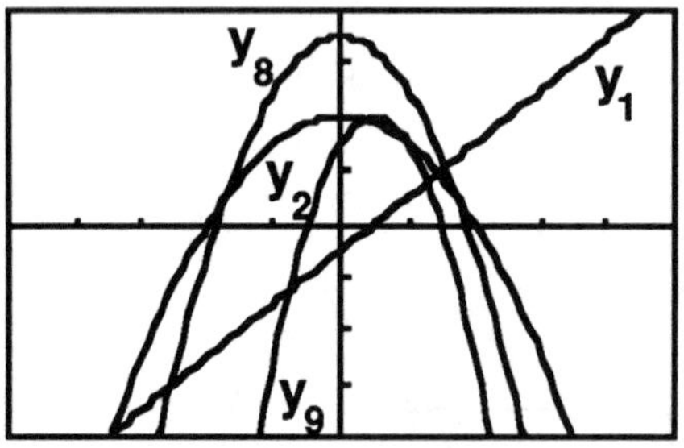

TI-81/82

Set the range to be: TI-81: [-9.6,9.4] x [-6.4,6.2], TI-82: [-9.4,9.4] x [-6.2,6.2] and in the **MODE** menu select TI-81: **Sequence**, TI-82: **Sequential**. In the Y variables list enter **2X–1** for **Y1**, **4–X²** for **Y2**, **Y1+Y2** for **Y3**, and **Y1–Y2** for **Y4**. The variable names **Y1** and **Y2** are obtained by selecting and entering the appropriate item from the **Y-VARS** menu. Upon graphing, the screen will appear as

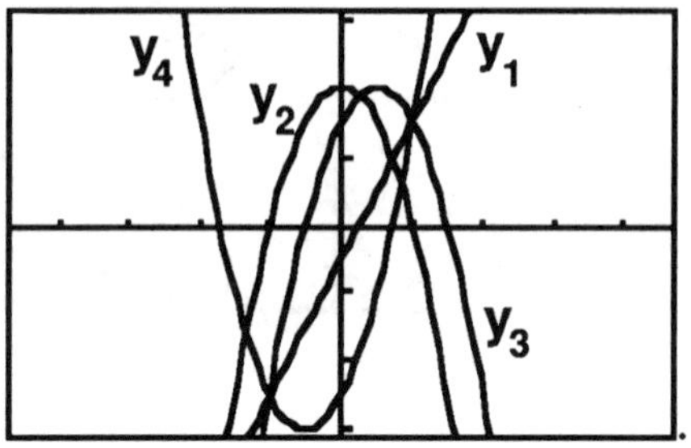

Reset the range to be [-5, 5] x [-10, 10] and keep **Y1** and **Y2** the same. Enter **Y1Y2** for **Y3** and unselect **Y4**. The graph of $y_5 = (fg)(x)$ will now be drawn. This looks like

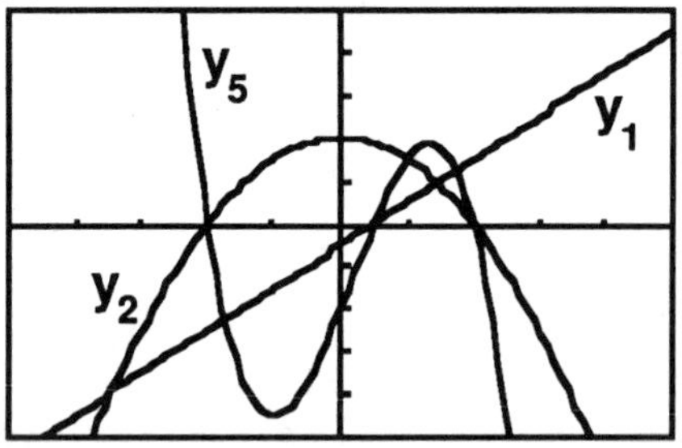

Repeating a similar procedure with **Y3** as **Y1/Y2**, which is the function $y_6 = (f/g)(x)$, and **Y4** as **Y1/Y3** which is the function $y_7 = (g/f)(x)$ and resetting the range to [-5, 5] x [-8, 8] we get the graphs shown at the top of the next page.

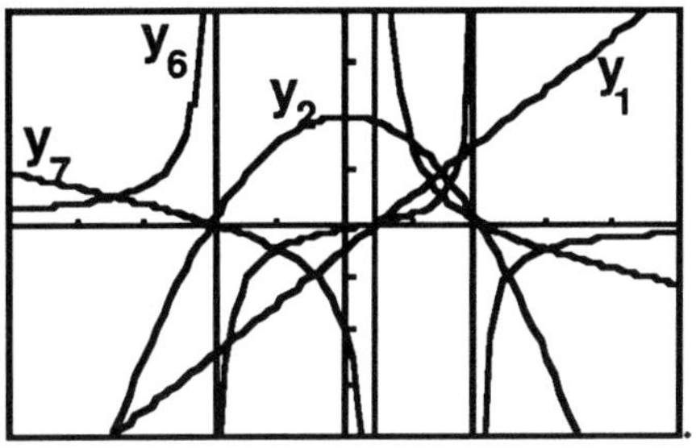

The vertical lines drawn through x = ±2 when **Y3** is drawn and the vertical line through x = 1/2 when **Y4** are not part of the graphs. The TI-81/82 will always do this with a graph of this type when the **Connected** option in the **MODE** menu is selected. If you redraw this graph with the **Dot** option selected, the vertical line is not drawn.

For the compositions, keep **Y1** and **Y2** as is and then enter for **Y3** the expression **2Y2–1** and for **Y4** enter the expression **4–Y1²**. These are, respectively, the functions

$$y_8 = f[g(x)] = 2(4 - x^2) - 1 = -2x^2 + 7$$

$$y_9 = g[f(x)] = 4 - (2x - 1)^2 = 4 - (4x^2 - 4x + 1) = -4x^2 + 4x + 3.$$

Drawing these graphs with the range [-6, 6] x [-4, 5] you should see

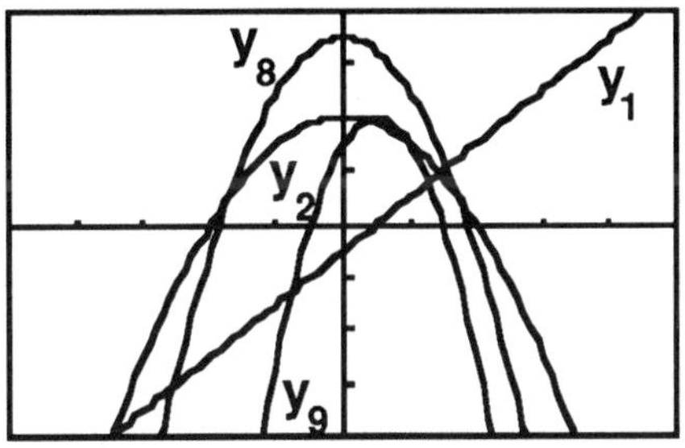

7.1 Exercises

For each function determine a, b, c and d so that af(b(x + c)) + d produces the given graph. In each case **Xscl** = **Yscl** = 1.

1. $f(x) = x^2$. (Note: Take b = 1)

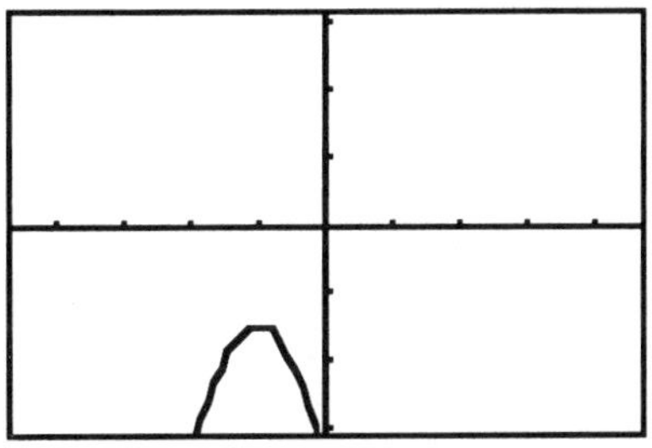

2. $f(x) = \frac{1}{x}$. (Note: Take b = 1)

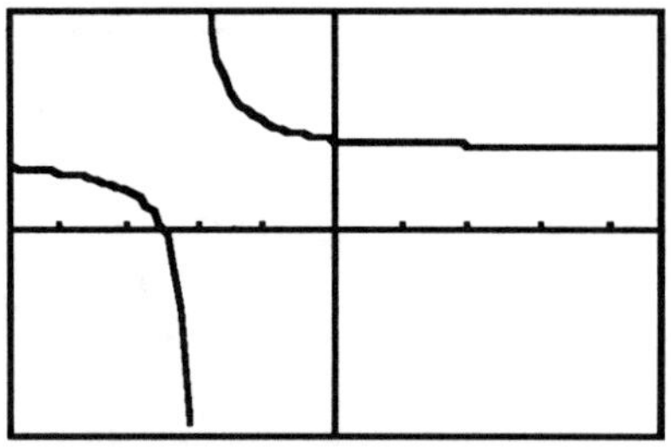

3. $f(x) = |x|$.

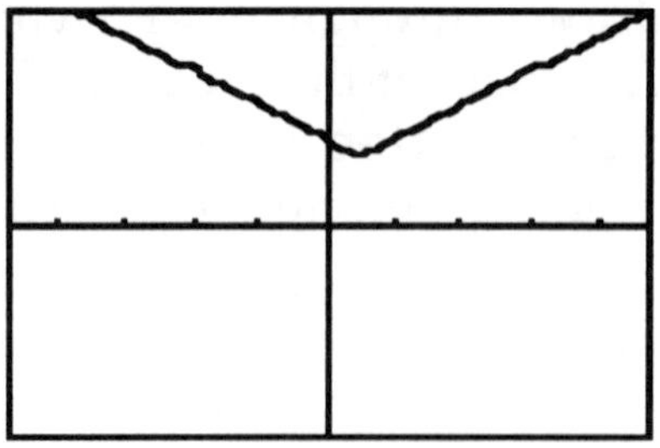

For each pair of functions draw the graphs of the sum, difference, product, quotients, and compositions. Read from the graph the domain of each.

4. $f(x) = 2x - 1, g(x) = \sqrt[3]{2 - x}$.

5. $f(x) = x^2 - 1, g(x) = \sqrt{x + 1}$.

6. $f(x) = \frac{2}{x - 3}, g(x) = x^2 - 4$.

7. $f(x) = x - x^2, g(x) = x + \frac{1}{x}$.

8. Let **V** denote the volume of a right circular cylinder and **S** its surface area. Suppose **R** = c**H**; then both **V** and **S** are functions of **H**. Draw the graphs of **V/S, S/V** and **V/RS**, for various values of c. What can you say about the graphs?

7.2 Tabular Functions

In many types of laboratory experiments or in statistical studies done in agriculture or business, there is a well defined set of input values, for example, grams of a reactant, pounds of fertilizer, or numbers of ball bearings produced. From these a set of values called the output or the response is obtained, for example, amount of product, yield of corn, or number of machinery repairs. In all of these situations an algebraic expression that relates the two sets of values may not be easily found, but if a graph of the input-output pairs is drawn it may be possible to at least construct an expression that relates the two sets of value with an acceptable amount of error.

• <u>Example 1:</u> A forester is concerned about the pattern of seed dispersal from trees. Knowledge of this pattern will aid in the understanding of the mechanisms and time of reforestation. So she counts the number of seeds found in concentric circles located at various distances from a tree. The data is given below.

Distance (feet) x	Number of Seeds y
2	33
4	17
6	13
8	10
10	7
12	5

Determine an approximate algebraic expression that relates x and y.

The first thing to do is plot the data. Using the range [0, 13] x [0, 34] the plot appears as

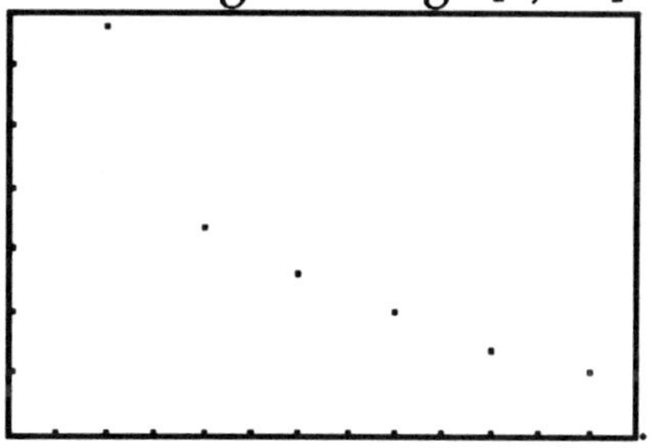

Looking at this plot, try to think of a function that has a similar look. After a little thought it seems to have the same shape as $f(x) = \frac{1}{x}$ but maybe translated and rescaled, i.e., maybe of the form

$$f(x) = \frac{a}{b(x + c)}.$$

This is the same form as

$$f(x) = \frac{a}{bx + c}.$$

where bc in the first form is the value of c in the second form. To see what the values of a, b, and c should be we will use a little algebra. We want

$$y = \frac{a}{bx + c}$$

or

$$\frac{a}{y} = bx + c$$

or

$$\frac{b}{a}x + \frac{c}{a} = \frac{1}{y}.$$

If we let X be the same as x, Y be $\frac{1}{y}$, and $M = \frac{b}{a}$, and $B = \frac{c}{a}$ we find that

$$Y = MX + B.$$

That is, $\frac{1}{y}$ is a linear function of x. This tells us that if y and x are related by the formula

$y = \frac{a}{bx + c}$ then the plot of x versus $\frac{1}{y}$ should be nearly a straight line. The pertinent data is in the following table.

Distance (feet) x	Reciprocal of Number of Seeds 1/y
2	0.0303
4	0.0588
6	0.0769
8	0.1000
10	0.1429
12	0.2000

Setting the range to be [0, 13] x [0, 0.22] and plotting data produces the graph

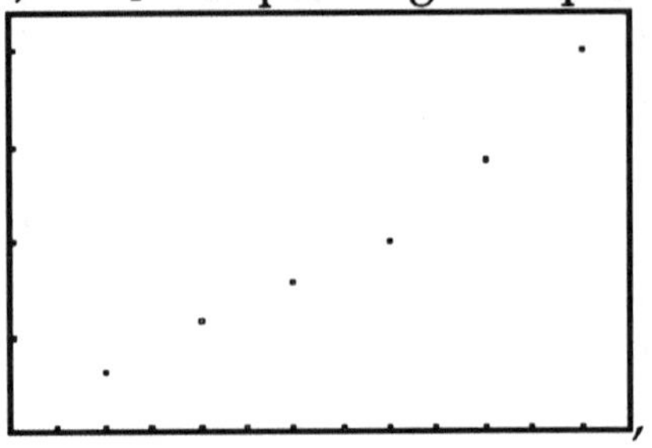

which is nearly a straight line. Taking $M = \frac{0.2 - 0.0303}{12 - 2} = 0.01697$ and choosing B so that the line passes through the averages of the x and 1/y values, which are 7 and 0.1015, respectively, we obtain the relation Y = 0.01697X – 0.01729. Drawing this line over the data plotted in the graph of above gives

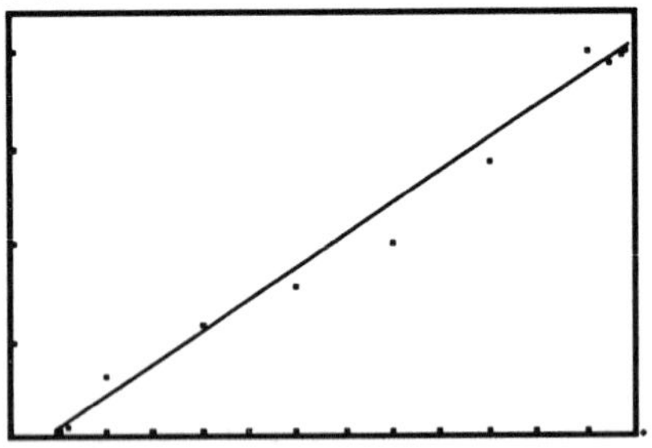

This appears to look pretty good. This relation tells us that

$$\frac{1}{y} = 0.01697x - 0.01729$$

or

$$y = \frac{1}{-0.01729 + 0.01697x}.$$

The table below gives a comparison between the number of seeds predicted by this equation and the number of seeds actually found.

Distance	Actual number of seeds	Predicted number of seeds
2	33	60
4	17	20
6	13	12
8	10	8
10	7	7
12	5	5

7.2 Exercises

1. The following table contains values for the volume, **V**, and surface area, **S**, of a pyramid with a square base and height **h** and the length of one side $\frac{5}{3}$ **h**. See the figure below.

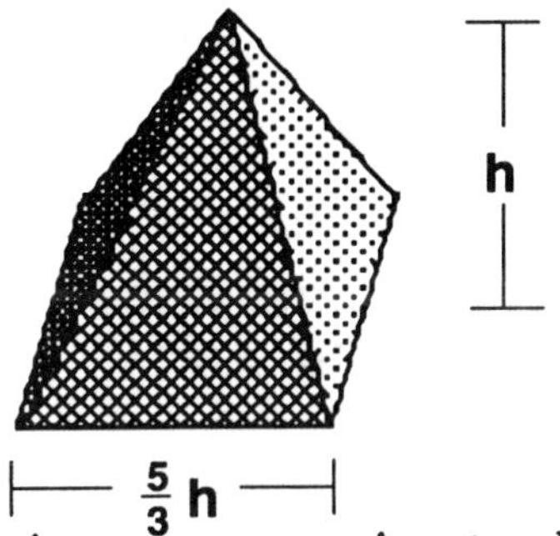

h	V	S
1.5	3.125	16.0128
3	25	64.0512
4.5	84.375	144.1153
6	200	256.2050
7.5	390.625	400.3203
9	675	576.4612
10.5	1071.875	784.6278
12	1600	1024.8200

Determine an approximate algebraic expression that relates **h** and **V** and **h** and **S**.Use these to determine how **V** and **S** are related. (Hint: Assume **S** = af(**h** + c) and **V** = ag(**h** + c) where f(x) and g(x) are of the form x^n for some positive integer n.)

2. An ellipse centered at the origin is determined by two positive numbers a > b > 0. See the figure below. The table below gives the area,A, and the perimeter, P, of an ellipse for a > 0 and b = $\frac{1}{2}$ a.

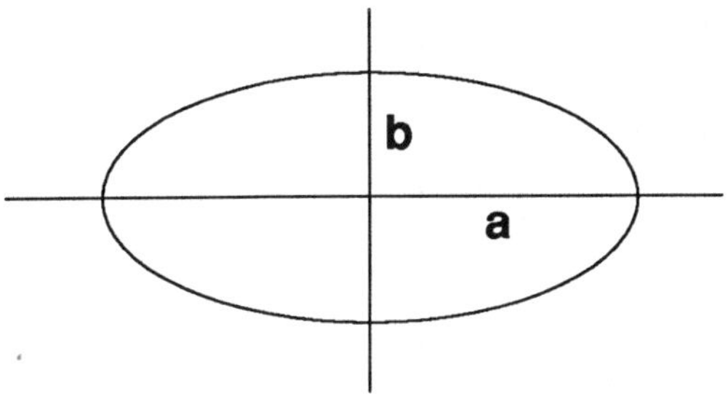

a	A	P
0.5	0.3927	2.4221
1	1.5707	4.8442
1.5	3.5343	7.2663
2	6.2832	9.6884
2.5	9.8175	12.1106
3	14.1372	14.5327
3.5	19.2423	16.9548
4	25.1327	19.3769
4.5	31.8086	21.7990
5	39.2699	24.2211

Determine approximate algebraic expressions relating a and A, a and P, and A and P.

3. A petroleum engineer is interested in determining the effective life of an oil well. He studies the oil production of a given well for a year and records the average number of barrels per day for a 12 month period. The data is as follows.

Month	Barrels per day
1	207
2	188
3	150
4	142
5	132
6	112
7	107
8	106
9	81
10	87
11	79
12	83

(a) Determine an algebraic expression relating the month and number of barrels per day.

(b) How many barrels per day does your expression predict for the 13th, 14th, and 15th month?

(c) The engineer knows that unless 45 barrels a day are pumped, the well will cease to be profitable. Based on this belief, what is the life of the well?

7.3 Inverse Functions

(McKeague: Section 9.4)

Suppose f is a given function. Is it always possible to find another function g such that

$$f[g(x)] = x \text{ and } g[f(x)] = x$$

for every value of x in both the domain of f and the domain of g? Such a function g is called an inverse of f. If a function is given by a table of values then an inverse relation can be obtained by exchanging the x and y values. For example if f is given by the table on the left, its inverse is given by the table on the right.

f

x	y
-3	3/11
-1.5	-3/13
0	0
.25	2/5
1.3	-13/19
2	-1/2
4.1	- 41/103

g

x	y
3/11	-3
-3/13	-1.5
0	0
2/5	.25
-13/19	1.3
-1/2	2
- 41/103	4.1

However, for the function given by the table

f						
x	-2	-1.5	-.5	0	.8	1.6
y	$-\frac{16}{3}$	-.2	$\frac{25}{3}$	4	4	$-\frac{4}{39}$

the inverse cannot be well defined for y = 4. The table given by

g						
x	$-\frac{16}{3}$	-.2	$\frac{25}{3}$	4	4	$-\frac{4}{39}$
y	-2	-1.5	-.5	0	.8	1.6

does give an inverse relation, but it is not a function.

This formal exchange of x and y values is how we will draw graphs of inverse functions and relations.

CASIO

We will discover that this program can also be used to draw functions that are defined in pieces.

Program 7.3.1C: Function Plotter

Keystrokes	Explanation
"A"?→A:"B"?→B ↵	Prompts for the input of the endpoints of the domain of the function.
(B–A)÷95→H ↵	Computes spacing for the evaluation of the function.
96→J:0→I ↵	Sets the number of times, **J**, a loop is executed. Sets the value of **I** that is used to compute the values of x where the function will be evaluated.
Lbl 1 ↵	Label 1 is the beginning of the evaluation loop.
A+IH→X ↵	Computes a value of x.
Prog 0 ↵	Calls program 0. This is where a function is evaluated and the value displayed or points are plotted. Any empty program number location can be used here. Program 0 will be used throughout this book.
Isz I ↵	Increments the value of **I** by one.
Dsz J ↵	Decrement **J** by one and if **J** is zero skip over the next command.
Goto 1 ↵	If **J** is not zero, go to label 1 and continue evaluation.

This program should be kept in memory since we will use it for many different applications.

• **Example 1** Given $f(x) = 2x^3 + x - 1$, draw the graph of its inverse.

Enter in program location 0, as specified in program 3.5.1C, the following commands.

```
2Xx^y3+X−1→Y ↵
Plot Y,X
```

These commands evaluate the function at the values of x computed in program 4.3.1C and then plot the pairs (y, x) which will produce a graph of the inverse relation for the function $f(x) = 2x^3 + x - 1$. Set the range to be the default and draw the graph of $y = 2x^3 + x - 1$. Which will appear as

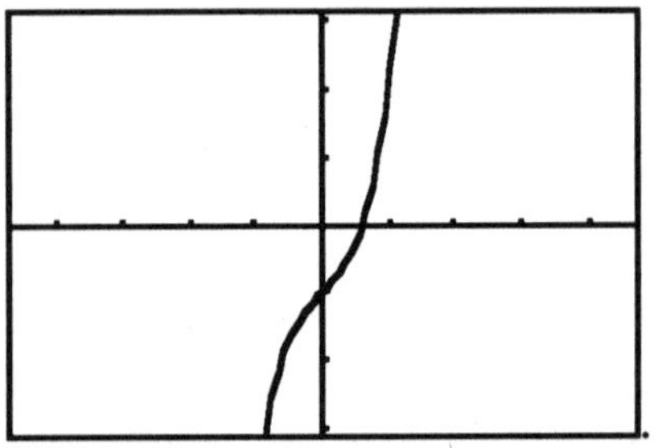

We see from the graph that f(x) is one-to-one and so its inverse is a well defined function. If we execute program 3.5.1C entering **A** = -4.7 and **B** = 4.7, we will see, through the flashing, the graph of the inverse being drawn. When complete, we should see

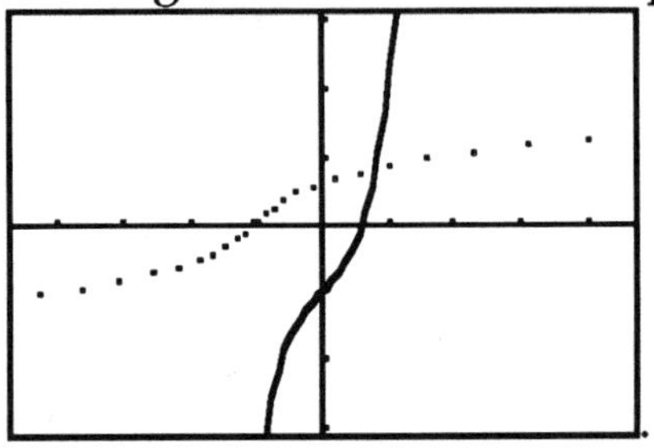

From this graph it is clear that f and its inverse are reflections through the line $y = x$.

It is certainly possible to draw the graph of an inverse function without the use of this program, provided the inverse can be found algebraically. If this is attempted with the above example, you will discover that it can be quite difficult or impossible.

• **Example 2:** Given $f(x) = x^2 - 2x$, $x \geq 1$, find its inverse and draw its graph.

Note that $f(x) = x^2 - 2x$ is only one-to-one if $x \geq 1$. If we set $y = x^2 - 2x$ and rearrange terms the quadratic equation

$$x^2 - 2x - y = 0$$

results with a = 1, b = -2, and c = -y. Using the quadratic formula to solve for x gives

$$x = 1 \pm \sqrt{1 + y}.$$

Which sign do we choose? Well since $x \geq 1$, we must take $x = 1 + \sqrt{1+y}$. Therefore the inverse of $f(x) = x^2 - 2x$ is $f^{-1}(x) = 1 + \sqrt{1+x}$. So drawing both $y = x^2 - 2x$ and $y = 1 + \sqrt{1+x}$ using the default range setting gives

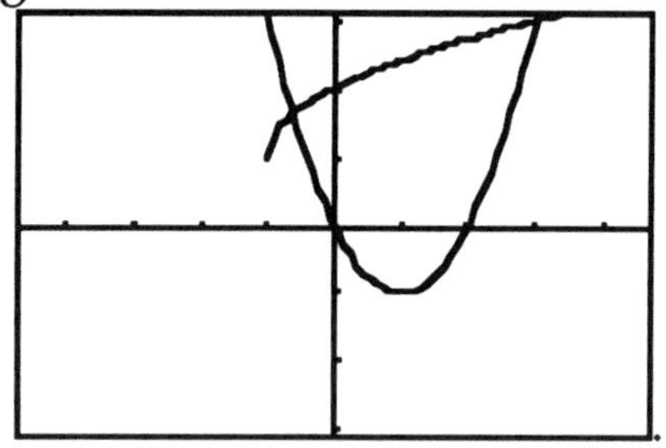

Using program 3.5.1C with

X²–2X→Y
Plot Y,X

in program location 0, produces

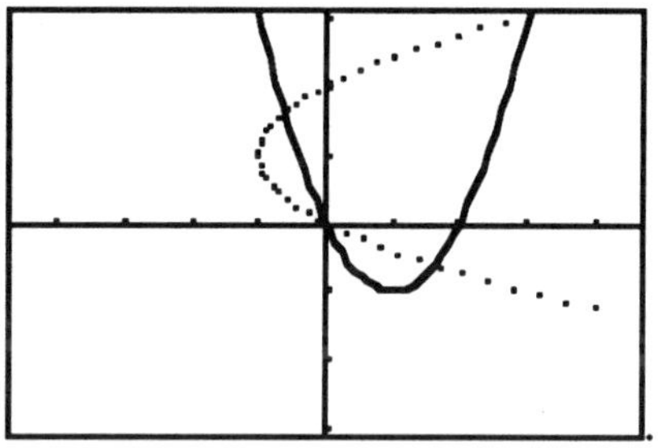

This is the complete inverse relation. Notice that the inverse relation is not a function since it fails the vertical line test.

- **Example 3:** Draw the graph of

$$f(x) = \begin{cases} -x - 3 \ ; \ x \leq -2 \\ x^2 - 3 \ ; \ -2 < x \leq 1 \\ \frac{1}{2}x - \frac{5}{2} \ ; \ x > 1 \end{cases}.$$

To produce this graph we will employ the point plotting program with the following program entered in program location 0.

Keystrokes	Explanation
X≤-2⇒Goto 1 ↵	Checks if $x \leq -2$; if so go to label 1 where the first part of the function is evaluated. If this fails then $x > -2$.
X≤1⇒Goto 2 ↵	Checks if $x \leq 1$; if so go to label 2 where the second part of the function is evaluated. If this fails then $x > 1$ and so the third part of the function is evaluated.
.5X-2.5→Y ↵	Here the third part of the function is evaluated.

Goto 3 ↵	Go to 3 where the graph is plotted.
Lbl 1 ↵	
-X–3→ Y↵	First part of the function is evaluated.
Goto 3 ↵	
Lbl 2 ↵	
X^2–3→Y ↵	Second part of the function is evaluated.
Lbl 3 ↵	
Plot X,Y	

Using the default range setting and entering -4.7 for **A** and 4.7 for **B** when executing program 7.3.1C produces

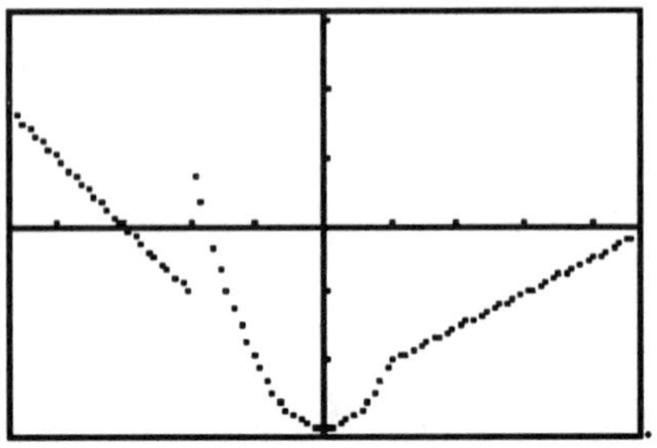

It is possible to draw the graph of this function on the fx-7700G without the program given above since it has the capability to draw piecewise functions. The following shows how this is done.

Graph Y=-X–3,[-4.7,-2]
:Graph Y=X^2–3,[-2,1]
:Graph Y=X÷2–5÷2,[1,4.7]

The graph produce will be identical to that above, except it will be connected. It is also possible to trace each piece, only that the trace will only work on one piece at a time. You must use the up or down arrow keys to switch to the other pieces.

TI-81

Program 7.3.1T: PTPLT, Point plotter and function evaluator

Commands	Explanation
:Disp "A"	Prompts for the input of **A**, the left endpoint.
:Input A	Input **A**.
:Disp "B"	Prompts for the input of **B**, the right endpoint.
:Input B	Input **B**.
:(B–A)/96→H	Computes the x increment.
:0→I	Sets the value of **I** to be 0.
:Lbl 1	Label 1 is the beginning of the loop.
:A+IH→X	Computes the value of x.
:PT-On(Y_1,X)	Plots the point (y, x), the inverse relation, where the function expression is entered in **Y_1** of the Y

	variables list.
:IS>(I,96)	Increments **I** by 1 and will skip the next command if **I** is greater than 96.
:Goto 1	

• <u>Example 1:</u> Given $f(x) = 2x^3 + x - 1$, draw the graph of its inverse.

In the Y variables list enter the expression **2X^3+X–1** and set the range to be the special default: TI-81: [-4.8, 4.7] x [-3.2, 3.1], TI-82: [-4.7, 4.7] x [-3.1, 3.1]. Execute the program **PTPLT** and enter -4.8 for **A** and 4.7 for **B**. The graph of $y = 2x^3 + x - 1$ will be drawn followed by a point plot of the inverse. The screen will appear as

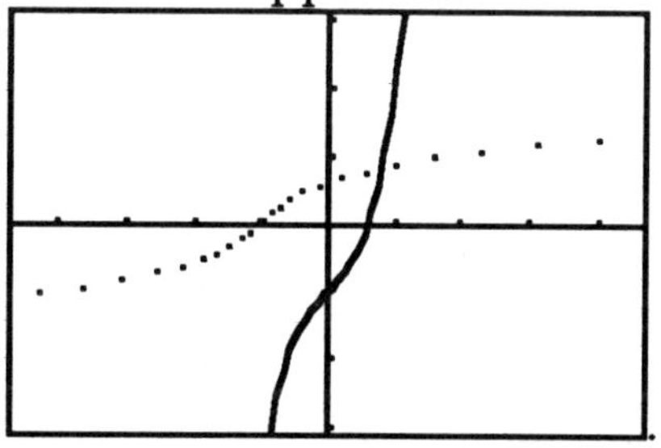

It is certainly possible to draw the graph of an inverse function without the use of this program provided the inverse can be found algebraically. If this is attempted with the above example, you will discover that it can be quite difficult or impossible.

• <u>Example 2:</u> Given $f(x) = x^2 - 2x$, $x \geq 1$, find its inverse and draw its graph.

Note that $f(x) = x^2 - 2x$ is only one-to-one if $x \geq 1$. If we set $y = x^2 - 2x$ and rearrange terms the quadratic equation

$$x^2 - 2x - y = 0$$

results with a = 1, b = -2, and c = -y. Using the quadratic formula to solve for x gives

$$x = 1 \pm \sqrt{1 + y}.$$

Which sign do we choose? Well since $x \geq 1$, we must take $x = 1 + \sqrt{1 + y}$. Therefore the inverse of $f(x) = x^2 - 2x$ is $f^{-1}(x) = 1 + \sqrt{1 + x}$. So drawing both $y = x^2 - 2x$ and $y = 1 + \sqrt{1 + x}$ using the special default range setting gives

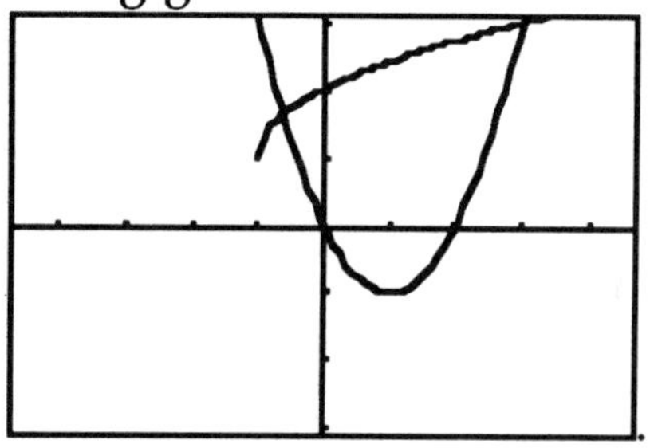

Using program 7.3.1T with **X²–2X** entered in **Y1** and unselecting all other expressions in the Y variables list produces the graph at the top of the next page.

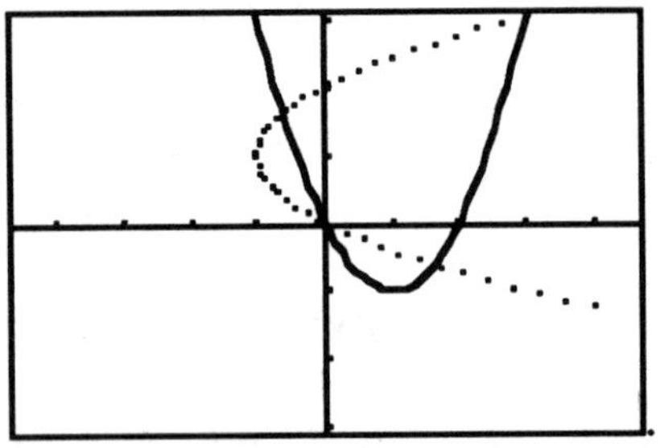

This is the complete inverse relation for $f(x) = x^2 - 2x$. Note that the inverse relation is not a function since it fails the vertical line test.

The TI-82 has the command **DrawInv**. This is the eighth item in the **DRAW** menu. We could use this to draw the inverse of the function $y = 2x^3 + x - 1$ by entering

DrawInv 2X^3+X–1.

The graph produced would be the same as that in example 1, except connected.

The TI-81/82 has the unique feature of being able to have logical tests as part of the expressions entered in the Y variables list. The syntax is (expression) (logical test). The logical test applies to the entire expression enclosed in parentheses. Two logical tests, each enclosed by parentheses, may follow an expression. If both tests are true the result is the expression times 1, if one test is false the result is the expression times 0. Groups of these expressions and logical tests can be added together to form piecewise defined functions.

• <u>Example 3:</u> Draw the graph of

$$f(x) = \begin{cases} -x - 3 \ ; \ x \le -2 \\ x^2 - 3 \ ; \ -2 < x \le 1 \\ \frac{1}{2}x - \frac{5}{2} \ ; \ x > 1 \end{cases}.$$

Enter the following for **Y1** in the Y variables list.

(-X–3)(X≤-2)+(X²–3)(-2<X)(X≤1)+(.5X–2.5)(X>1).

Set the range to be the special default, set the draw mode to be **Dot** and draw the graph. The graph should appear as follows

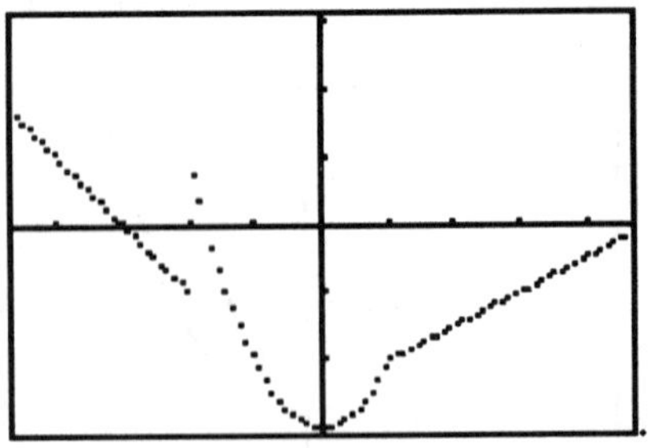

Because it is drawn from the Y variables list it is possible to use the trace just as you would with any other function.

Caution: **If a function like $y = \sqrt{1 - x^2}$ is used in a piecewise defined function, because it is not defined for $|x| > 1$, it can cause the pieces of the function defined for $|x| > 1$ to not be drawn. (This is not the situation on the TI-82.)**

7.3 Exercises

Determine where f(x) is one-to-one, find its inverse algebraically for each function, and draw the graph of each.

1. $f(x) = -5x + 2.$
2. $f(x) = \sqrt[3]{2x - 5}.$
3. $f(x) = \dfrac{4x}{x + 2}.$
4. $f(x) = -x^2 + 4x.$
5. Draw the graph of $f(x) = \begin{cases} \frac{1}{2}x \ ; \ x < -2 \\ \dfrac{-4x}{x^2 + 4} \ ; \ -2 \le x < 2 \\ -2x + 3 \ ; \ x > 2 \end{cases}$. Determine if f(x) is one-to-one and if so draw the graph of $f^{-1}(x)$.

Draw the graph of the following functions.

6. $f(x) = \begin{cases} 2x - 1 \ ; \ x < 1 \\ 1 \ ; \ x \ge 1 \end{cases}.$
7. $f(x) = \begin{cases} -1 - x \ ; \ x < -1 \\ 1 - x^2 \ ; \ -1 \le x < 1. \\ x - 1 \ ; \ x \ge 1 \end{cases}$
8. $f(x) = \begin{cases} 0 \ ; \ x < -\frac{3}{2} \\ \frac{1}{2}\left(x + \frac{3}{2}\right)^2 \ ; \ -\frac{3}{2} \le x < -\frac{1}{2} \\ -x^2 + \frac{3}{4} \ ; \ -\frac{1}{2} < x \le \frac{1}{2} \\ \frac{1}{2}\left(\frac{3}{2} - x\right)^2 \ ; \ \frac{1}{2} < x \le \frac{3}{2} \\ 0 \ ; \ x > \frac{3}{2} \end{cases}.$

7.4 The Circle

(McKeague: Section 9.5)

A circle is defined as the set of points in the plane that are equidistant from a fixed point. It is then necessary to look first at how distance can be found between two points in the plan The formula for finding the distance between two points (x_1, y_1) and (x_2, y_2) is $d = \sqrt{(x_1 - x_2)^2 + (y_1 - y_2)^2}$. The following programs will take points from the screen and find their distance.

CASIO

Program 7.4.1C: Distance Formula

Keystrokes	Explanation
Cls ↵	Clears the graphics screen.
Range -47,47,10,-31,31,10 ↵	Sets the range.
Plot ◢	Sets a point at the screen center to be moved. Program halts. The screen coordinate is automatically stored in memory locations **X** and **Y**.
X→S:Y→T ↵	Stores the point in **S** and **T**.
Plot ◢	Sets a point at the screen center to be moved. Program halts.
X→U:Y→V ↵	Stores the second point in **U** and **V**.
$\sqrt{((S-U)^2+(T-V)^2)}\to D$ **↵**	Computes the distance and stores the value in **D**.
"D":D ◢	Displays the distance.
Line	Draws a line between the two points.

TI-81/82

Program 7.4.1T: DISTF Distance Formula

Commands	Explanation
:ClrDraw	Clears the graphics screen.
:All-Off (FnOff)	Unselects all of the expressions in the Y variables list. Found in the **Y-VARS** menu.
:-48→Xmin	The next 6 lines set the range. The variables **Xmin**, **Xmax**, **Xscl**, **Ymin**, **Ymax**, and **Yscl** are found in the **VARS** menu under the **RNG** title.
:47→Xmax	
:10→Xscl	
:-32→Ymin	
:31→Ymax	
:10→Yscl	
:Input	This activates the graphics screen with movable crosshairs. The screen coordinate is automatically stored in memory locations **X** and **Y**.
:X→S	Stores the first point in **S** and **T**.
:Y→T	
:PT-On(S,T)	Plots the first point.
:Input	This activates the graphics screen with movable crosshairs.

:X→U	Stores the second point in **U** and **V**.
:Y→V	
:√((S−U)²+(T−V)²)→D	Computes the distance and stores the value in **D**.
:Disp "D"	
:Disp D	Displays distance.
:Pause	This command suspends the program. To continue press [ENTER]. Found in the **CTL** menu when on line 2 or greater of an edit screen of a program.
:Line(S,T,U,V)	Draws a line between the two points.

• <u>Example 1:</u> Find the distance between the points (-1, 8) and (5, 12)

Execute the appropriate program given above and move the dot or crosshairs to the point (-1, 8) and press [EXE] or [ENTER]. This puts -1 into location **S** and 8 into location **T**, plots the point, and sets another dot or crosshairs at the screen center. Move to the point (5, 12) and press [EXE] or [ENTER]. The distance should be **D = 7.211102551**. Pressing [EXE] or [ENTER] one more time will draw a line between the points.

The programs can be changed to accept keyboard input as follows.

CASIO
Replace the first six lines with the following

```
"X1"?→S:"Y1"?→T ↵
"X2"?→U:"Y2"?→V ↵
```

and delete the last line.

TI-81
Replace the first 15 lines with the following

```
:Disp "X1"
:Input S
:Disp "Y1"
:Input T
:Disp "X2"
:Input U
:Disp "Y2"
:Input V
```

and delete the last two lines.

The equation of a circle of radius r and centered at the point (a, b) is $(x - a)^2 + (y - b)^2 = r^2$. Solving this equation for y gives the two equations

$$y = \sqrt{r^2 - (x - a)^2} + b \qquad y = -\sqrt{r^2 - (x - a)^2} + b$$

The first equation describes the upper half circle and the second equation describes the lower half circle. The following programs will take a given point as the center and radius and draw the circle.

CASIO

Program 7.4.2C: Draws Circles

Keystrokes	Explanation
"A"?→A:"B"?→B ↵	Prompts for input of center.
"R"?→R ↵	Prompts for input of radius.
Plot A,B ↵	Plots the center point.
Graph Y=√(R²–(X–A)²)+B ↵	Draws the upper half of the circle.
Graph Y=-√(R²–(X–A)²)+B	Draws the lower half of the circle.

TI-81/82

Program 3.1.3T: CIRCLES Draws Circles

Commands	Explanation
:All-Off (FnOff)	Unselects all expressions in Y variables list.
:Disp "A"	Prompts for the input of the x-coordinate of center.
:Input A	
:Disp "B"	Prompts for the input of the y-coordinate of center.
:Input B	
:Disp "R"	Prompts for the input of the radius.
:Input R	
:"√(R²–(X–A)²)+B"→Y1	Creates expression defining the upper half and stores in **Y1**.
:"-√(R²–(X–A)²)+B"→Y2	Creates expression defining the lower half and stores in **Y2**.
:Y1-On	
:Y2–On	
:DispGraph	Displays the graphics screen causing the circle to be drawn.
:Pt-On(A,B)	Plots the center point.

When drawing circles, in order for them to look right the screen must be given a square coordinate system. For example, the default setting on the CASIO or the default setting on the TI-81 are both square coordinate systems. The utility programs 2.1.1C and 2.1.1T will do this automatically.

• <u>Example 2:</u> Find the equation of the circle with center (-3, 2) and radius 5. Draw the graph. (McKeague: Example 3, Section 9.5)

In this case a = -3, b = 2, and r = 4 so the equation is $(x + 3)^2 + (y - 2)^2 = 25$. Set the range to be

CASIO: [-14.75, 8.75] x [-5.75, 9.75], **Xscl** = **Yscl** = 2.5
TI-81: [-15, 8.75] x [-6, 9.75], **Xscl** = **Yscl** = 2.5
TI-82: [-14.75, 8.75] x [-5.75, 9.75], **Xscl** = **Yscl** = 2.5.

These range settings were set using the programs 3.1.2C and 3.1.2T with **X** = 3, **Y** = -5 and **S** = .25. The following graph is then produced.

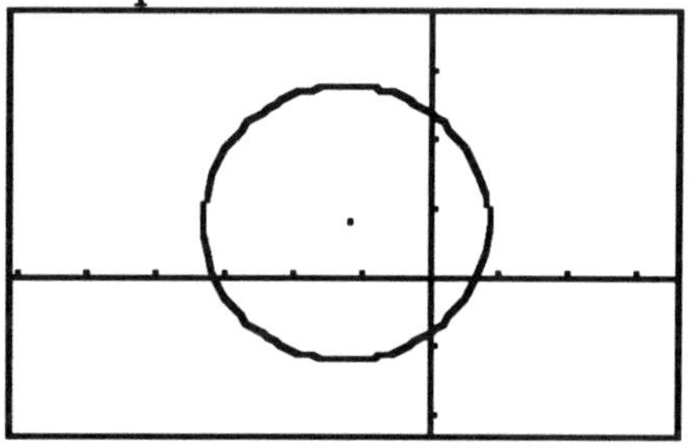

• <u>Example 3:</u> Find the center and radius of the circle whose equation is

$$4x^2 + 4y^2 - 16x - 4y - 3 = 0$$

and draw the graph.

By completing the square we have

$$4(x^2 - 4x + 4) + 4(y^2 - y + \tfrac{1}{4}) - 3 - 16 - 1 = 0$$

or

$$4(x - 2)^2 + 4\left(y^2 - \frac{1}{2}\right)^2 = 20$$

or

$$(x - 2)^2 + \left(y - \frac{1}{2}\right)^2 = 5.$$

So the center is $\left(2, \frac{1}{2}\right)$ and the radius is $\sqrt{5}$. The graph, using the default range setting, looks like

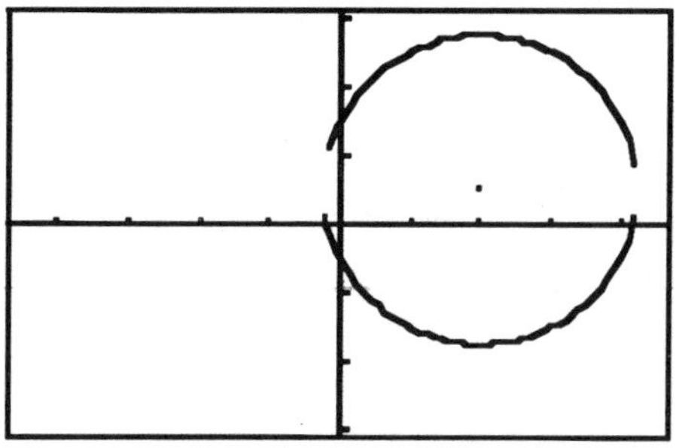

Notice that the graph is broken. This is because of the inability to obtain an exact value for $\sqrt{5}$.

As written, program 7.4.1T cannot draw more than one circle on the screen at a time, but it is possible to use the trace. In order to draw more than one circle on the screen at a time, but losing the ability to trace, simply replace lines 8 and 9 with

```
:DrawF √(R²–(X–H)²)+K
:DrawF -√(R²–(X–H)²)+K.
```

7.4 Exercises

Find the distance between the given points and draw a line joining the points.

1. (-5, 1), (4, 8).
2. (-3, 3), (-8, -1).
3. $(\sqrt{2}, 0), (1,1)$.
4. $\left(\frac{3}{5}, \frac{4}{5}\right), \left(-\frac{4}{5}, \frac{3}{5}\right)$.

Find the center and radius of the circle with the given equation and draw its graph.

5. $x^2 + y^2 + 2x - 4y + 1 = 0$.
6 $16x^2 + 16y^2 - 12x + 32y = 23$.
7. $x^2 + y^2 - 6x + 4y + 5 = 0$.

8. With the default settings; CASIO: [-4.7, 4.7] x [-3.1, 3.1], TI-81: [-4.8, 4.7] x [-3.2, 3.1], TI-82: [-4.7, 4.7] x [-3.1, 3.1] draw the graphs of the circles $(x - 2)^2 + (y - 2)^2 = \frac{1}{4}$ and $(x - 2)^2 + (y - 2)^2 = \frac{1}{16}$, but smile.

7.5 Ellipses and Hyperbolas

(McKeague: Section 9.6)

Ellipse

The ellipse with foci P and Q is the set of points X = (x, y) such that

distance from X to P + distance from X to Q = r.

The figure below illustrates this definition.

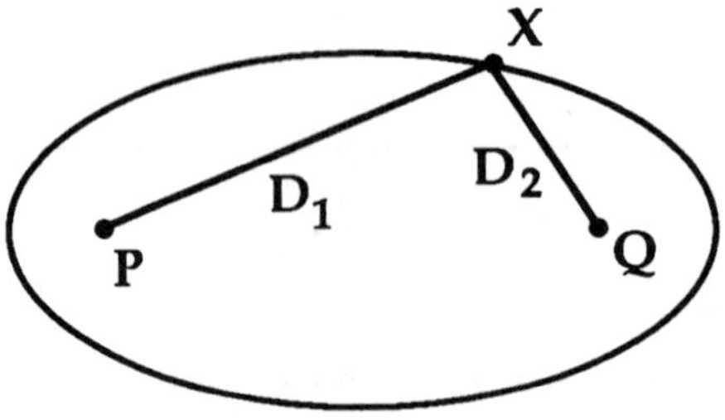

$D_1 + D_2 = r$

The midpoint of the line segment joining P and Q is called the center. If the coordinates of the center are (x_0, y_0) then the *standard form* of the equation of the ellipse is given by

$$\frac{(x-x_0)^2}{a^2}+\frac{(y-y_0)^2}{b^2}=1.$$

If $a > b$ then the foci are $(x_0 - c, y_0)$ and $(x_0 + c, y_0)$ where $c^2 = a^2 - b^2$. If $b > a$ the the foci are $(x_0, y_0 - c)$ and $(x_0, y_0 + c)$ where $c^2 = b^2 - a^2$. The vertices are given by $(x_0 \pm a, y_0)$ and $(x_0, y_0 \pm b)$. The following graphs illustrate these two cases.

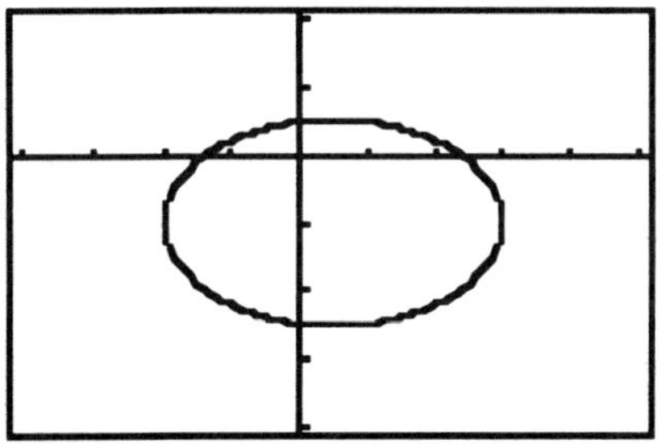

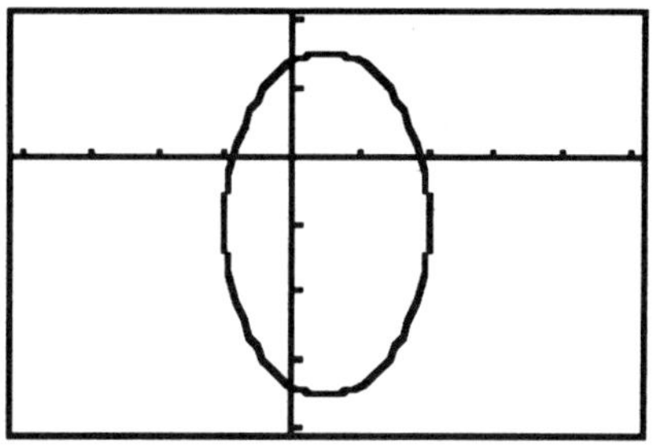

In either case we solve the equation for y and draw the graphs of the resulting functions. Producing these graphs with the calculators is accomplished using the following techniques.

CASIO

The program that follows is the easiest way to produce the graph of an ellipse whose equation is given in the standard form.

Keystrokes	Explanation
"A"?→A:"B"?→B ↵	Prompts for the input of the coefficients **A** and **B**.
"X0"?→H:"Y0"?→K ↵	Prompts for the input of the center **X0** and **Y0**.
Prog _ ↵	The blank should be replaced with the program location number that contains **RANGE** program given by program 2.1.1C in chapter 2. When this program is executed enter **H** for **X** and **K** for **Y**. The value of 0.2 is usually good choice for **S**.
Graph Y=B√(A^2–(X–H)^2)÷A+K ↵	Draws the upper half of the curve.
Graph Y=-B√(A^2–(X–H)^2)÷A+K ↵	Draws the lower half of the curve.

TI-81/82

In the Y variables list enter for **Y1** the expression **B√(A^2–(X–H)^2)÷A+K** and enter for **Y2** the expression **-B√(A^2–(X–H)^2)÷A+K**. Store the values of **A**, **B**, **H** and **K** in those memory locations. The values of **H** and **K** are the coordinates of the center of the ellipse. To produce the correctly scaled graph execute the **RANGE** program (program 2.1.1T in chapter 2) and enter **H** for **X**, **K** for **Y** and the value of 0.2 for **S**; this value seems to be pretty good most of the time. Now simply draw the graphs of each of the expressions stored in the Y variables list.

The ratio

$$e = \begin{cases} \frac{c}{a} \ ; \ a > b, c^2 = a^2 - b^2 \\ \frac{c}{b} \ ; \ b > a, c^2 = b^2 - a^2 \end{cases}$$

is called the eccentricity of an ellipse and is a number between 0 and 1. If e = 0 then the ellipse is a circle, i.e., a = b. If e is nearly 1 the ellipse is very flat. The graphs below show ellipses with various eccentricities.

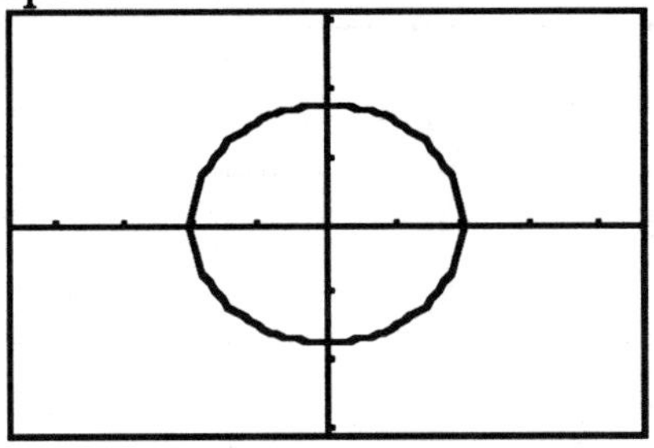

$e = 0.5, a^2 = 1, b^2 = 3/2$

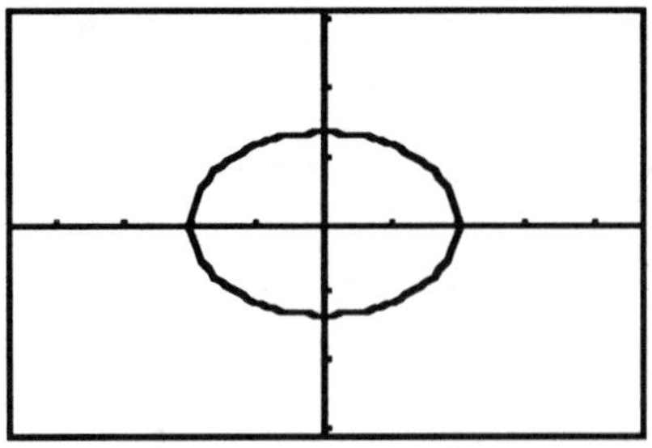

$e = 0.75, a^2 = 1, b^2 = 7/16$

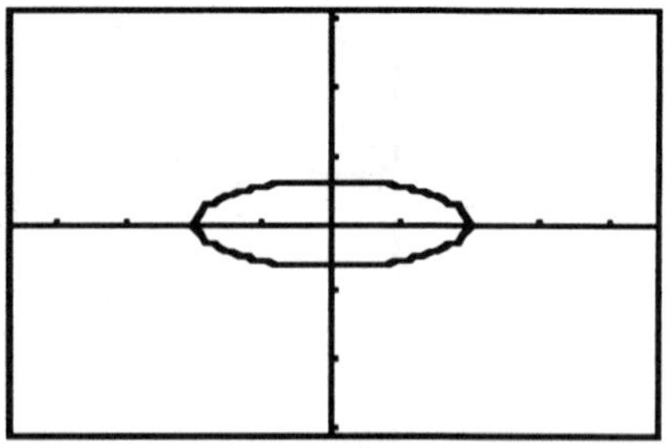

$e = 0.95, a^2 = 1, b^2 = 39/400$

These graphs were produced by taking c = e and a = 1 and then taking $b = \sqrt{a^2 - c^2}$. For example if e = 0.5 then $b = \sqrt{1 - 0.5^2} = \sqrt{\frac{3}{4}} = \frac{\sqrt{3}}{2}$.

• **Example 1:** Put the following equation of an ellipse in standard form and find the center and foci of the curve. Draw the graph.

$$x^2 + 2y^2 + 8y = 0.$$

In order to determine the standard form of each of these equations it is necessary to complete the square for both the x and y terms.

Completing the square produces the sequence of expressions

$$x^2 + 2(y^2 + 4y \quad) = 0$$

$$x^2 + 2(y^2 + 4y + 4) = 0 + 2(4)$$

$$x^2 + 2(y + 2)^2 = 8$$

$$\frac{x^2}{8} + \frac{(y + 2)^2}{4} = 1.$$

From the standard form we see that the center of the ellipse is (0, -2), $a = 2\sqrt{2}$, b = 2 and so $c = \sqrt{8 - 4} = 2$. Therefore the foci are (2, -2) and (-2, -2) and $e = \frac{1}{\sqrt{2}} = 0.7071067814$. Using the technique described above, enter **A** = $2\sqrt{2}$, **B** = 2, **X0** = 0, and **Y0** = -1. When the **RANGE** program executes enter **X** = **H**, **Y** = **K**, and **S** = 0.125. The graph at the top of the next page will be produced. The foci are also plotted. This was done after the graph of the ellipse is drawn. Notice that the graph is not closed

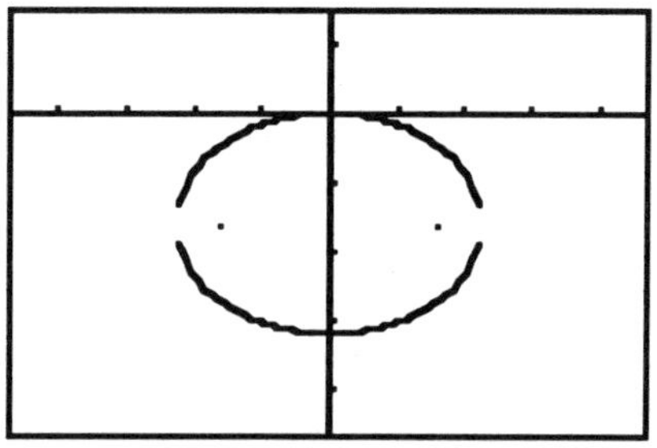

because of the precision of the calculator.

- **Example 2:** A family of ellipses have eccentricity e, a focus at (0, 0) and a major axis having length 4 along the x-axis. Find the equation of this family depending on e and draw several of the graphs of this family.

Since the length of the major axis is 4 we have that $e = \frac{c}{2}$. So we will take c = 2e. This implies that $b^2 = a^2 - c^2 = 4(1 - e^2)$. Since the major axis is along the x-axis, the center has coordinates (h, 0) where h could be positive or negative. Since it doesn't make any difference which, let's take h positive. Now the foci are given by (h ± c, 0) and one of them is (0, 0). So because we are taking h to be positive we must have the center to the right of the focus (0, 0). Therefore (h – c, 0) = (0, 0) and so h = c = 2e. Putting this together gives the equation

$$\frac{(x - 2e)^2}{4} + \frac{y^2}{4(1 - e^2)} = 1.$$

Solving for y gives the two equations

$$y = \sqrt{1 - e^2}\sqrt{4 - (x - 2e)^2}, \quad y = -\sqrt{1 - e^2}\sqrt{4 - (x - 2e)^2}.$$

To draw the graphs of several of these ellipses the following programs are useful.

CASIO

```
Cls:Lbl 1
"E"?→E
E=0⇒Goto 2
Graph  Y=√(1–E²)√(4–(X–2E)²)
Graph  Y = -√(1–E²)√(4–(X–2E)²)
Goto 1
Lbl 2
```

TI-81/82

```
ClrDraw
Lbl 1
Disp "E"
Input E
If E=0
Goto 2
DrawF  √(1–E²)√(4–(X–2E)²)
DrawF  -√(1–E²)√(4–(X–2E)²)
Pause
Goto 1
Lbl 2
Stop
```

The test to see if **E** = 0 is simply used to stop the program. Because using the Y variables list does not allow you to draw more than 4 graphs on the screen the **DrawF** command is used in the TI-81 program. The following figure shows ellipses for e = 0.2, 0.4, 0.6, and 0.8.

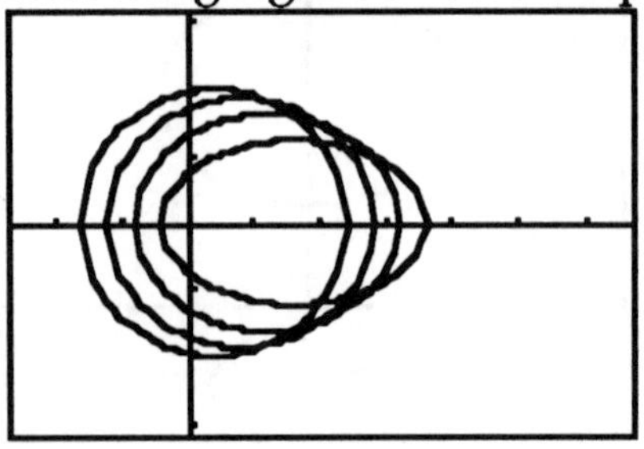

The range for this graph was set using the **RANGE** program with **X** = 2, **Y** = 0, and **S** = 0.1.

Hyperbolas

The hyperbola with foci P and Q is the set of points X = (x, y) such that

| distance from X to P – distance from X to Q | = r.

The figure below illustrates this definition.

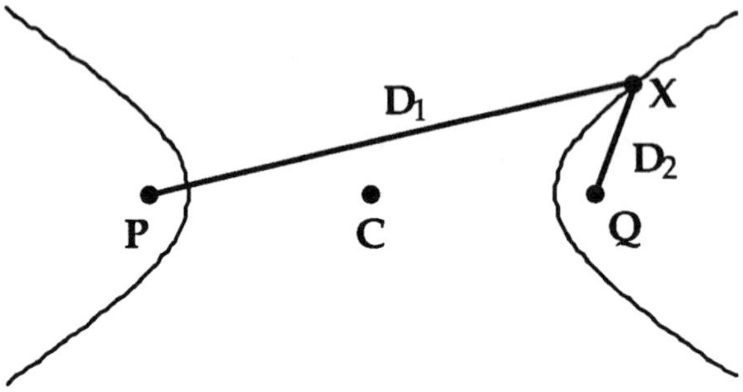

$|D_1 - D_2| = r$

The midpoint of the line segment joining P and Q is called the center. If the coordinates of the center are (x_0, y_0) then there are two possible *standard forms* for the equation of the hyperbola.

Form 1

$$\frac{(x-x_0)^2}{a^2} - \frac{(y-y_0)^2}{b^2} = 1$$

Vertices: $(x_0 \pm a, y_0)$

Foci: $(x_0 \pm c, y_0)$, $c^2 = a^2 + b^2$

Asymptotes: $y = \pm \frac{b}{a}(x - x_0) + y_0$

To graph: draw the functions $f_1(x) = \frac{b}{a}\sqrt{-a^2 + (x-x_0)^2} + y_0$, $f_2(x) = -\frac{b}{a}\sqrt{-a^2 + (x-x_0)^2} + y_0$.

Form 2

$$\frac{(y-y_0)^2}{b^2} - \frac{(x-x_0)^2}{a^2} = 1$$

Vertices: $(x_0, y_0 \pm b)$

Foci: $(x_0, y_0 \pm c)$, $c^2 = a^2 + b^2$

Asymptotes: $y = \pm\frac{b}{a}(x - x_0) + y_0$

To graph: draw the functions $f_1(x) = \frac{b}{a}\sqrt{a^2 + (x-x_0)^2} + y_0$, $f_2(x) = -\frac{b}{a}\sqrt{a^2 + (x-x_0)^2} + y_0$.

These two cases are illustrated in the figures below. The center and foci are plotted on each graph as well.

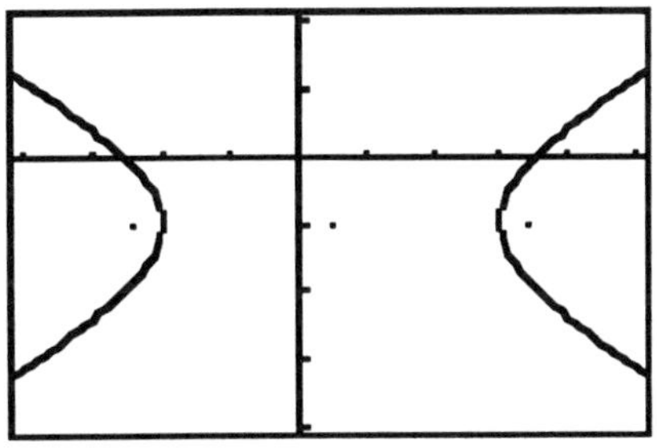

$$\frac{(x-1)^2}{25} - \frac{(y+2)^2}{9} = 1$$

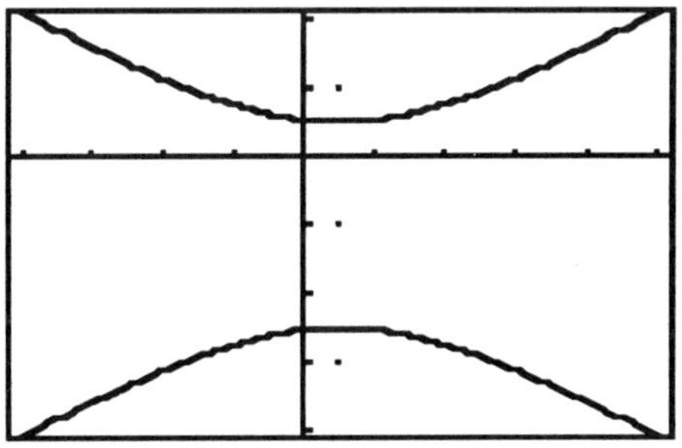

$$\frac{(y+2)^2}{9} - \frac{(x-1)^2}{25} = 1$$

In either case the following procedures can be used to draw the graph of a hyperbola.

CASIO

The program that follows is the easiest way to produce the graph of a hyperbola whose equation is given in the standard form.

Keystrokes	Explanation
"A"?→A:"B"?→B ↵	Prompts for the input of the coefficients **A** and **B**.
"FORM 1 OR 2"?→F ↵	Prompts for **F**, to determine if the hyperbola is form 1 or form 2.
F=1⇒-1→U ↵	If **F** = 1, set **U** = -1, graph $y = \pm\frac{b}{a}\sqrt{-a^2 + (x-x_0)^2} + y_0$
F=2⇒1→U ↵	If **F** = 2, Set **U** = 1, graph $y = \pm\frac{b}{a}\sqrt{a^2 + (x-x_0)^2} + y_0$.
"X0"?→H:"Y0"?→K ↵	Prompts for the input of the center **H** and **K**.
Prog _ ↵	The blank should be replaced with the program location number that contains **RANGE** program given by program 2.1.1C in chapter 2. When this program is executed enter **H** for **X** and **K** for **Y**. The value of 0.2 is usually good for **S**.
Graph Y=B√(UA²+(X −H)²)÷A+K ↵	Draws the upper half of the curve.

Graph Y=-B√(UA²+(X–H)²)÷A+K ↵ Draws the lower half of the curve.

TI-81/82

In the Y variables list enter for **Y1** the expression **B√(-A²+(X–H)²)÷A+K** and enter for **Y2** the expression **-B√(-A²+(X–H)²)÷A+K**. Store the values of **A**, **B**, **H** and **K** in those memory locations. The values of **H** and **K** are the coordinates of the center (x_0, y_0). To produce the correctly scaled graph execute the **RANGE** program (program 3.1.2T in chapter 3) and enter **H** for **X**, **K** for **Y** and the value of 0.2 for **S**; this value seems to be pretty good most of the time. Now simply draw the graph of each of the expressions stored in the Y variables list. This will produce the graph of a hyperbola if the standard form is form 1. If the standard form is form 2 enter for **Y1** the expression **B√(A²+(X–H)²)÷A+K** and enter for **Y2** the expression **-B√(A²+(X–H)²)÷A+K**.

The ratio

$$e = \begin{cases} \dfrac{c}{a}\ ; & \text{if the equation is } \dfrac{(x-h)^2}{a^2} - \dfrac{(y-k)^2}{b^2} = 1 \\[2ex] \dfrac{c}{b}\ ; & \text{if the equation is } \dfrac{(y-k)^2}{b^2} - \dfrac{(x-h)^2}{a^2} = 1 \end{cases}$$

is called the eccentricity of the hyperbola. Note that since $c^2 = a^2 + b^2 > a^2$ and $c^2 = a^2 + b^2 > b^2$ we have $e > 1$. If e is near 1 the branches of the hyperbola are very steep and if e is large the branches are very flat. The graphs below show hyperbolas with various eccentricities. All of these graphs are given by the equation $\frac{y^2}{b^2} - \frac{x^2}{a^2} = 1$.

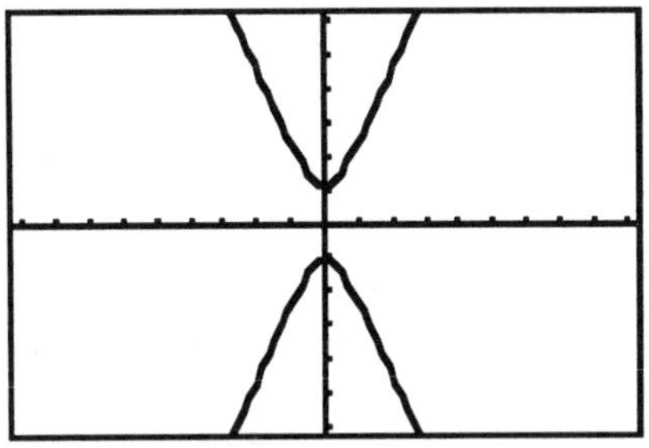

$e = 1.1, b^2 = 4, a^2 = 0.84$

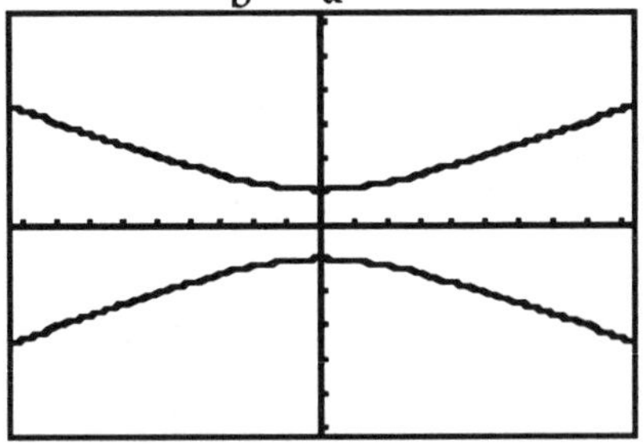

$e = 3, b^2 = 4, a^2 = 32$

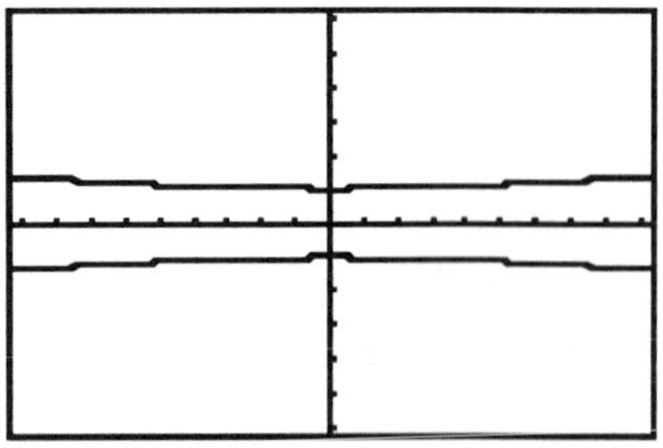

$e = 10, b^2 = 4, a^2 = 396$

• <u>Example 1:</u> Put the following equation of a hyperbola in standard form and find the center, foci, vertices and eccentricity of each curve. Draw the graph.

$$-3x^2 + 4y^2 + 18x + 8y - 35 = 0.$$

Rewrite the equation as $-3(x^2 - 6x \quad) + 4(y^2 + 2y \quad) = 35$. Completing the square on both the x and y terms produces

$$-3(x^2 - 6x + 9) + 4(y^2 + 2y + 1) = 35 - 27 + 4$$
$$-3(x-3)^2 + 4(y+1)^2 = 12$$
$$\frac{(y+1)^2}{3} - \frac{(x-3)^2}{4} = 1.$$

The center is therefore (3, -1). Now $a^2 = 4$, $b^2 = 3$ so $c = \sqrt{4+3} = \sqrt{7}$ and hence the eccentricity is $\sqrt{\frac{7}{3}}$. The foci are then $(3, -1 \pm \sqrt{7})$ and the vertices are $(3, -1 \pm \sqrt{7})$. Solving the equation for y produces the two equations

$$y = \frac{\sqrt{3}}{2}\sqrt{4 + (x-3)^2} - \text{ and } y = -\frac{\sqrt{3}}{2}\sqrt{4 + (x-3)^2} - 11.$$

The asymptotes are given by the equations $y = \pm\frac{\sqrt{3}}{2}(x-3) - 1$.

CASIO

Execute the program given above entering **A** = 2, **B** = $\sqrt{3}$, **F** = 2 for form 2, **X0** = 3 and **Y0** = -1. When the **RANGE** program executes enter **H** for **X**, **K** for **Y** and 0.2 for **S**. To draw the asymptotes enter **Graph Y=√3(X–3)÷2–1** and **Graph Y=-√3(X–3)÷2–1** after the hyperbola is drawn.

TI-81

Enter in the Y variables list **√3√(4+(X–3)²)/2–1** for **Y1** and **-√3√(4+(X–3)²)/2–1** for **Y2**. Execute the **RANGE** program and enter 3 for **X**, -1 for **Y** and 0.2 for **S** and then draw the graphs of **Y1** and **Y2**. To draw the asymptotes enter **√3(X–3)÷2–1** for **Y3** and **-√3(X–3)÷2–1** for **Y4**.

Either procedure produces the following graph. The foci and asymptotes are also plotted.

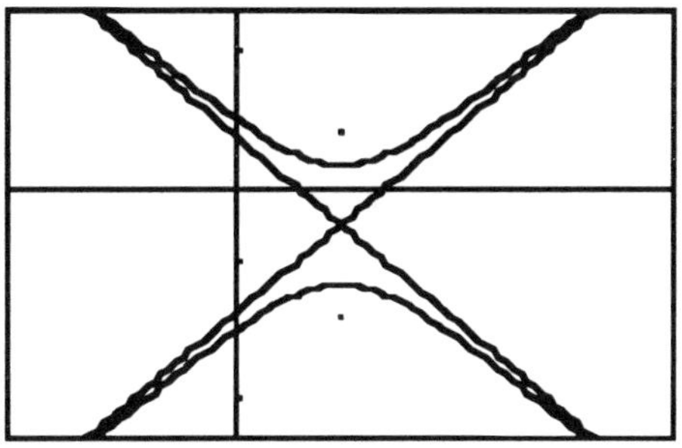

7.5 Exercises

Write the equation of the ellipse in standard form. Find the center, foci, and eccentricity. Draw the graph of each.

1. $9x^2 + 4y^2 - 18x - 27 = 0.$
2. $3x^2 + 5y^2 + 12x - 20y + 17 = 0.$
3. $4x^2 + 16y^2 + 4x + 96y + 141 = 0.$
4. $3x^2 + 8y^2 + 24x - 32y + 56 = 0.$

5. The following table gives the eccentricities and lengths of the semi-major axis of the planets and Halley's Comet. Using the equation in exercise 8 and the programs given in example 8.1.2 draw the graphs of the orbits of the first 4 planets and Halley's comet on the same graph and then draw the graphs of the orbits of the last 5 planets and Halley's comet on the same graph. Set the range using the **RANGE** programs 3.1.2C or 3.1.2T on pages 60 and 61 of chapter 3. Answer the following questions about the orbits. Does Halley's Comet come closer to the sun than the planet Venus? Is the planet Pluto always the furthest planet from the sun?

Planet	Eccentricity	Semi-major axis
Mercury	0.206	35.9†
Venus	0.007	67.1
Earth	0.017	92.6
Mars	0.093	141.7
Jupiter	0.048	483.4
Saturn	0.054	888.6
Uranus	0.046	1783.3
Neptune	0.008	2796.2
Pluto	0.248	3672.3
Halley's Comet	0.968	1671.4

(† millions of miles)

Write the equation of the hyperbolas in standard form. Find the center, foci, and eccentricity. Draw the graph of each.

6. $9x^2 - 4y^2 - 18x + 8 = 0.$

7. $x^2 - 4y^2 + 2x + 16y - 14 = 0$

8. $4x^2 - 3y^2 - 8x + 6y - 11 = 0.$

9. $-2x^2 + y^2 + 6y + 25 = 0.$

8. A family of hyperbolas have eccentricity e, a focus at (0, 0), a = 3, and axis of symmetry along the y axis. Determine a equation that describes the family and draw the graphs of several.

7.6 Nonlinear Systems

(McKeague: Section 9.7)

Just as with linear systems the solution of the nonlinear systems are points where the graphs intersect.

• <u>Example 1:</u> Find all solutions of the system

$$\frac{x^2}{9} + \frac{y^2}{4} = 1$$
$$x^2 + y^2 = 5.$$

The solution must be a point where the graphs

$$y = \frac{2}{3}\sqrt{9 - x^2} \text{ or } y = -\frac{2}{3}\sqrt{9 - x^2} \text{ and } y = \sqrt{5 - x^2} \text{ or } y = -\sqrt{5 - x^2}$$

intersect. These expressions are obtained by solving each of the equations for y. Graphing these functions using the default range setting produces the following graph.

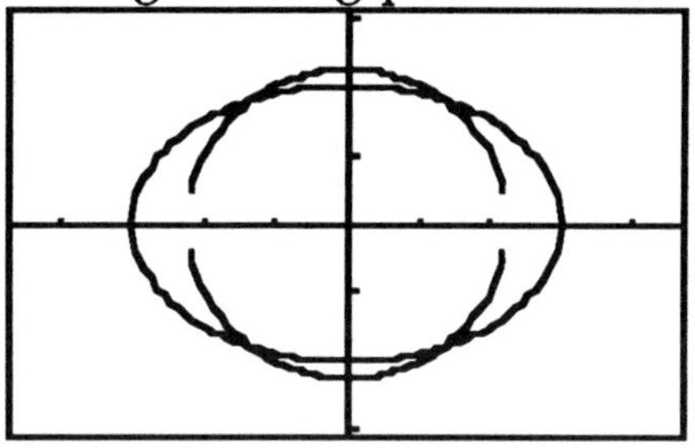

We see from the graph that the ellipse $\frac{x^2}{9} + \frac{y^2}{4} = 1$ and the circle $x^2 + y^2 = 5$ intersect four times. Zooming in on the intersection point in quadrant I three times using a factor of 10, we find x = 1.342 and y = 1.789. Checking the range we find that **Xmax** – **Xmin** = 0.01 and **Ymax** – **Ymin** = 0.01. This tells that this approximate solution is accurate to about one decimal place. Substituting these values into the equation we find

$$(1.342)^2/9 + (1.789)^2/4 = 1.0002 \text{ and } (1.342)^2 + (1.789)^2 = 5.0015.$$

Not too bad. If we take $y^2 = 5 - x^2$ and substitute this into the equation of the ellipse we have

$$\frac{x^2}{9} + \frac{5 - x^2}{4} = 1$$

$$4x^2 + 9(5 - x^2) = 36$$

$$-5x^2 = 36 - 45.$$

So we must have $x^2 = 9/5$ or $x = \pm\frac{3}{\sqrt{5}} = \pm\frac{3}{5}\sqrt{5}$. This then implies that $y^2 = 16/5$ or $y = \pm\frac{4}{\sqrt{5}} = \pm\frac{4}{5}\sqrt{5}$.

7.6 Exercises

Find approximate solutions of each system by zooming in on the intersection points of the graphs until the error is 0.001, i.e., until **Xmax** – **Xmin** = 0.001 and **Ymin** – **Ymin** = 0.001. Substitute these approximate solutions into the equations to see how well they satisfy the system. Use the substitution method to find the exact solutions.

1. $x + y = 8$
   $xy = 4.$
2. $x^2 + y^2 = 4$
   $x^2 - y^2 = 2.$
3. $x^2 - y^2 = 4$
   $x - 2y = 3.$
4. $4x^2 + y = 1$
   $x^2 - 2y = 0.$

Investigating Inverses

Parametric equations can be used to graph relations and their inverses. What is a parametric equation? If the value of the x- coordinate of a graph is expressed as a function of a variable t and if the y coordinate of the same graph is expressed as a function of the same variable t, then the t is called a parameter. The resulting equations describing the graph are called parametric equations. The following equations represent a parabola in parametric form.

$$x = t \qquad y = t^2$$

To graph parametric equations the mode of the calculator must be changed from function grapher to parametric graph (Figure 1). When set in this mode the ranges for x, y and the parameter t can be adjusted (Figure 2). All the figures in this investigation were generated with a TI-82. The TI-82 is similar. On the CASIO fx-7700G use [SHIFT] [MODE] [X]. This is PRM mode or parametric graphing mode.

Figure 1

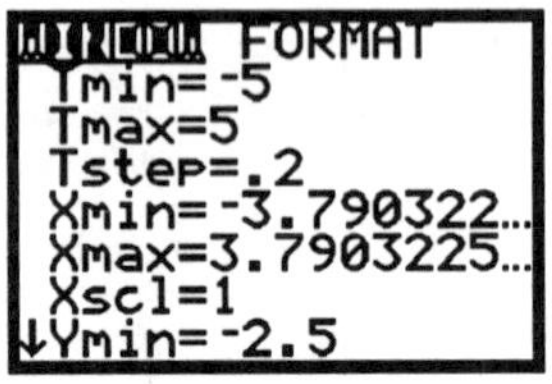

Figure 2

The equation screen (y =) has two equations for each graph, X_{1T} and Y_{1T}. Figure 3 shows the equations necessary to graph the parabola, $y = x^2$. Figure 4 shows the graph. On the CASIO fx-7700G enter **Graph(X,Y)=(T,T²)**. Set the range to the default setting and on the second range screen set **min** = -5, **max** = 5, and **ptch** = .2 and draw the graph. It will look similar to Figure 4.

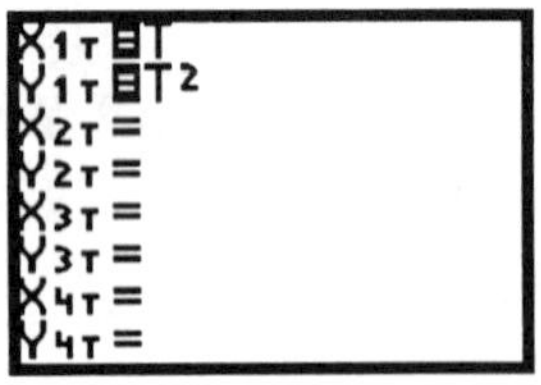

Figure 3

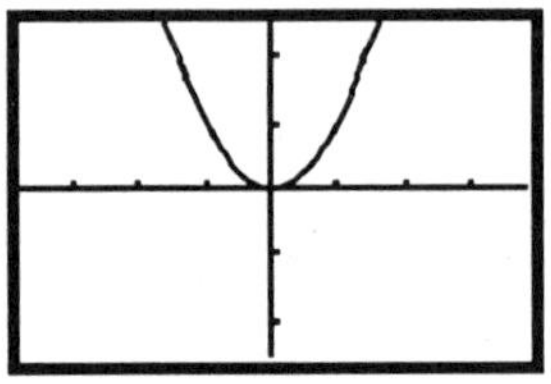
Figure 4

The inverse of a relation interchanges the x and y variables. Explain how to do this in the equation screen so that the inverse is graphed. The resulting graphs, $y = x^2$ *and* $x = y^2$*, are illustrated in Figure 5. Also, investigate how to include the line* $y = x$ *on the graph so it is apparent the inverse is the reflection of the original graph (Figure 6).*

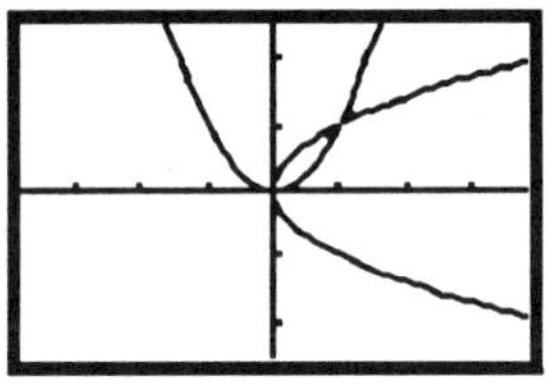
Figure 5

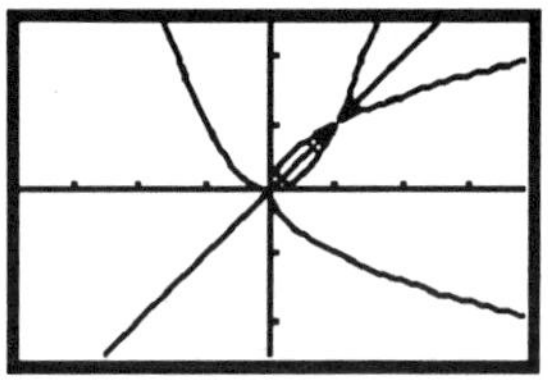
Figure 6

Graph each relation and its inverse. Determine if the original curve is a function and if the inverse is a function. Complete Table 1 to assist you in the investigation.

Original Equation	Is the original equation a function?	Is the inverse of the original equation a function?
$y = x^2 + 2$		
$y = x^4 - 2x^3$		
$y = x^2 + 2x - 3$		
$y = x^3 - 2$		
$y = 4x - 1$		
$y = x^5 - 2x^3 - 2$		
$y = 2^x$		
$y = \left(\frac{1}{2}\right)^x$		
$x = y^2 - 3$		
$x = y^3 - 3y$		

Table 1

From your investigation discuss how to determine if an equation is a function and if its inverse is a function. Can a relation which is not a function have an inverse which is a function?

Chapter 8

Exponential and Logarithmic Functions

Exponential and logarithmic functions are used to describe a great number of processes in economics, biology and engineering. These functions are quite simple to describe but difficult to evaluate except by a calculator or computer. The graphics calculator can thus be used to full advantage to evaluate and graph these functions.

8.1 Exponential Functions

(McKeague: Section 9.3)

Let a be a positive number. The function $f(x) = a^x$ is called an exponential function. The domain of this function is the set of all real numbers, and we will see that since the base is positive the range is the set of all positive real numbers. The shapes of the graphs of these functions for different values of a are all quite similar; the major difference occurs when the base a is between 0 and 1.

• **Example 1:** Compare the graphs of $f(x) = 2^x$, $g(x) = \left(\frac{7}{2}\right)^x$, and $h(x) = \left(\frac{1}{3}\right)^x$. How do th functions differ for x near 0 and for $|x|$ large?

Note that $f(0) = g(0) = h(0) = 1$, so all of these functions have the same y-intercept. The graphs of these functions are given in the figures below. The graph on the left has the range setting is $[-5, 5] \times [-5, 30]$. From the graph on the left we see that both f(x) and g(x) are

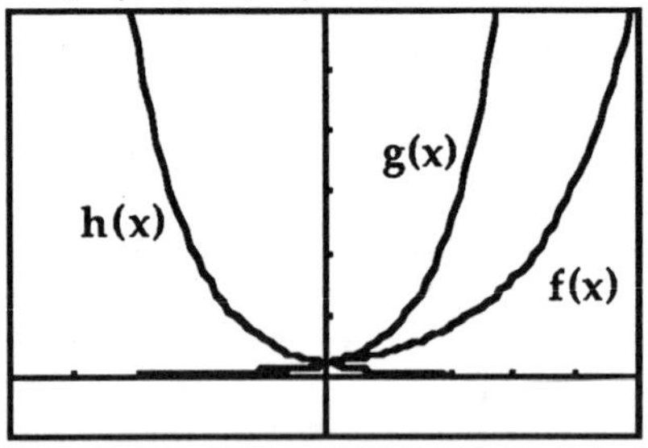

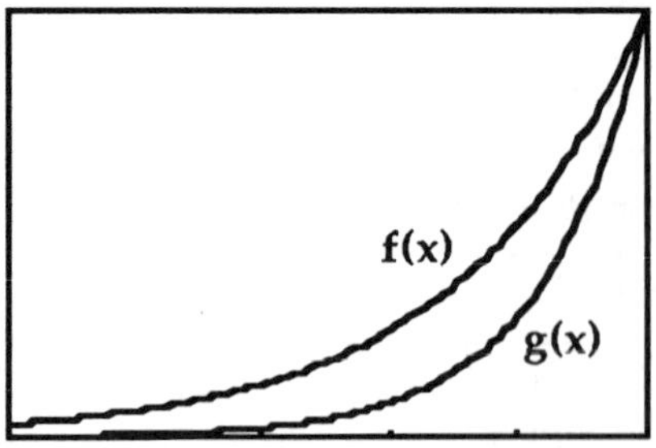

increasing but h(x) is decreasing. This is because the base for the function h(x) is less than 1. The graphs of f(x) and g(x) illustrate exponential growth while h(x) illustrates exponential decay. It appears that if $x > 0$, $2^x < \left(\frac{7}{2}\right)^x$ but what about when $x < 0$? Note that when $x < 0$ both f(x) and g(x) are below $y = 1$ so if we redraw the graphs of the range $[-5, 0] \times [0, 1]$, as shown above right, then we reveal that $2^x > \left(\frac{7}{2}\right)^x$ if $x < 0$. It is also apparent that as x approaches $-\infty$ both f(x) and g(x) approach zero but h(x) is unbounded. Also as x approaches $+\infty$ both f(x) and g(x) are unbounded, but h(x) approaches zero.

• **Example 2:** An amount P is deposited in a savings account paying an annual interest rate r compounded quarterly. The amount in the account at any time t, in years, is given by the function

$$A(t) = P\left(1 + \frac{r}{4}\right)^{4t}$$

Suppose P = \$2000.00 and r = 0.075 (7.5%). How long does it take for the original amount to double? How long will it take to become a millionaire?

The function governing the growth of the money in the savings account is

$$A(t) = 2000\left(1 + \frac{.075}{4}\right)^{4t}$$
$$=2000(1.01875)^{4t}.$$

To determine how long it will take to double we must find the value of t where A(t) = 4000, i.e., the point where the graph of A(t) and the graph of y = 4000 intersect. Using the range [0, 20] x [2000, 5000] we get the graph

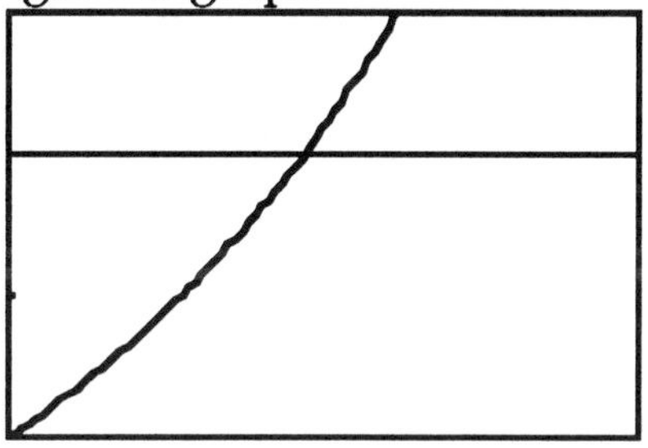

By zooming in we find that t = 9.3283. So it takes approximately 9 years and 4 months for the initial investment to double. To find how long it will take to become a millionaire find where the graphs of A(t) and y = 1000000 intersect. Drawing these graphs on the range [0, 100] x [2000, 1500000] produces the graph

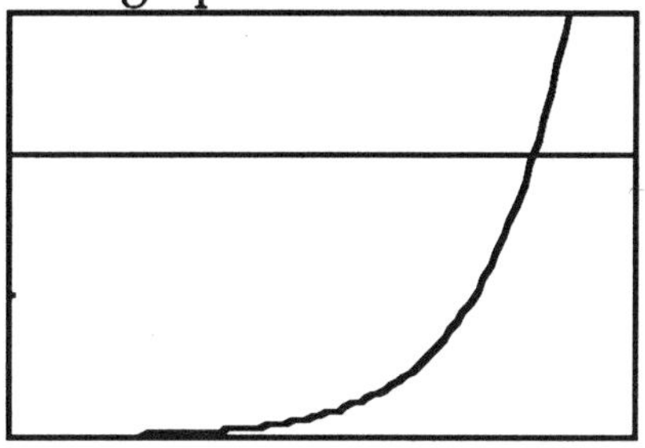

Zooming in on the point of intersection we find t = 83.6359. It will therefore take about 83 years and 8 months to become a millionaire this way.

A different type of exponential function where the base varies is the function $f(x) = \left(1 + \frac{1}{x}\right)^x$, x > 0. The graph of this function using the range [0, 20] x [1, 3] is given at the top of the next page.

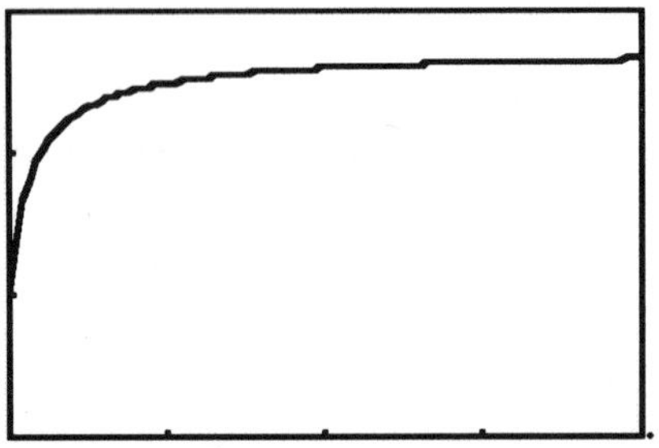

It appears that as x goes to $+\infty$ this function has a horizontal asymptote. The following table illustrates this. The underlined digits are the digits that are correct.

x	f(x)
10^2	2.70481382942
10^3	2.71692393224
10^4	2.71814592683
10^5	2.71826823717
10^6	2.71828046932
10^7	2.71828169254
10^8	2.71828181487
10^9	2.71828182710
10^{10}	2.71828182832

The actual value rounded to 30 decimal places is 2.718281828459045235360287471353. This number has properties similar to those of π in that it is irrational and has an infinite non-repeating decimal expansion. This number is denoted by e and is used as the base of the "natural" exponential function $f(x) = e^x$. On the calculators this function is evaluated by

CASIO: [SHIFT] [e^x] (number)

TI-81/82: [2nd] [e^x] (number)

where (number) is a specific value. For example on the screen e^3 would appear as

CASIO: **e 3**

TI-81/82: **e^3.**

Drawing the graphs of $g(x) = 2^x$, $f(x) = e^x$, and $h(x) = 3^x$ we see that f(x) is between g(x) and h(x) since $2 < e < 3$.

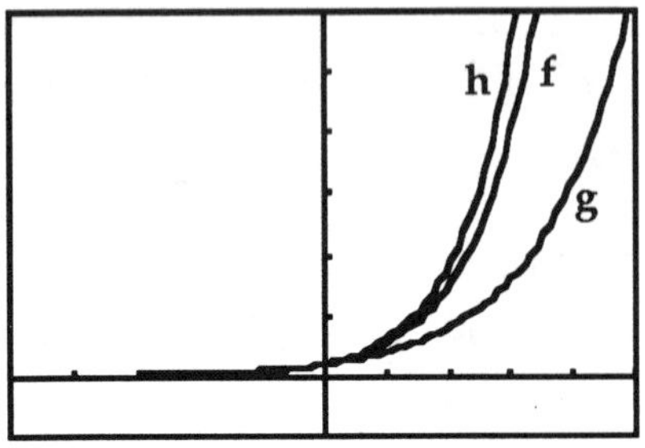

Range: [-5, 5] x [-5, 30]

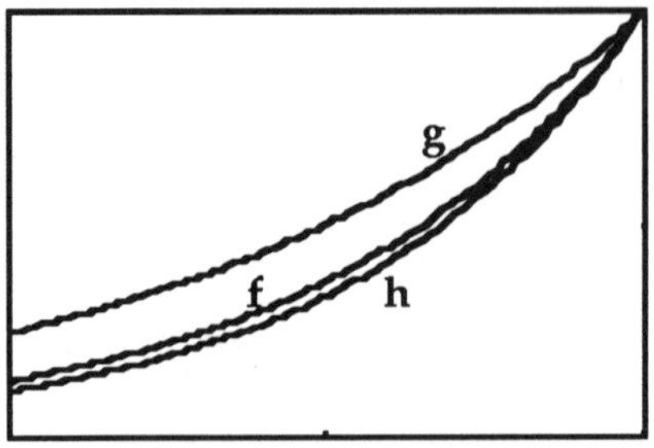

Range: [-2, 0] x [0, 1]

It is this natural exponential function that is used to model radioactive decay and population growth. In the study of radioactive substances the interest lies in the time required for half of original amount of the substance to decay. This time is called the half-life of the substance. In population models the interest is in the time required for the population to double its size.

- **Example 3:** Carbon-14, ^{14}C, is a radioactive isotope of carbon. Suppose initially we have 2 grams of ^{14}C. Then the amount q(t) at any time t, in years, is given by the function

$$q(t) = 2e^{-0.00012422t}.$$

Find the half-life of ^{14}C?

The function q(t) is a function that illustrates exponential decay. If we use a little algebra we find

$$\begin{aligned} q(t) &= 2e^{-0.00012422t} \\ &= 2(e^{-0.00012422})^t \\ &= 2(0.999876^t). \end{aligned}$$

Because the base is less than one we know that q(t) is a decreasing function. To determine the half-life of ^{14}C we must solve the equation q(t) = 1, but this is the point where the graphs of q(t) and y = 1 intersect. Drawing these two graphs with the range setting [0, 10000] x [0, 2] produces

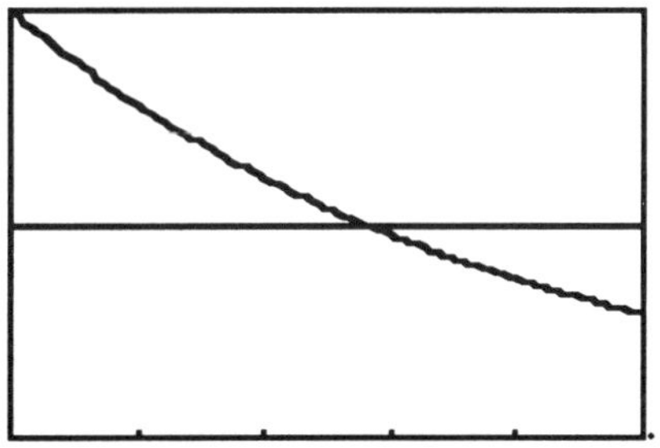

Zooming in on the intersection point we find that t = 5580. Therefore the half-life of C^{14} is 5580 years.

- **Example 3 (continued):** ^{14}C is used to date ancient objects that were once part of a living organism. The half-life of ^{14}C tells us that 5580 years after an organism dies it contains half as much ^{14}C as it did when alive. Suppose a fossil is unearthed and it is determined that the amount of ^{14}C present is 35% of what it would be for a similarly sized living organism. What is the approximate age of the fossil?

Suppose q_0 is the orginal amount of ^{14}C. The we know that

$$q(t) = q_0e^{-0.00012422t}.$$

If t is the number of years since death then by assumption $q(t) = 0.35q_0$. Thus

$$0.35 = \frac{q(t)}{q_0} = e^{-0.00012422t}.$$

So we must find where the graphs of $y = e^{-0.00012422t}$ and $y = 0.35$ intersect. Drawing these graphs using the range [0, 10000] x [0, 0.7] reveals

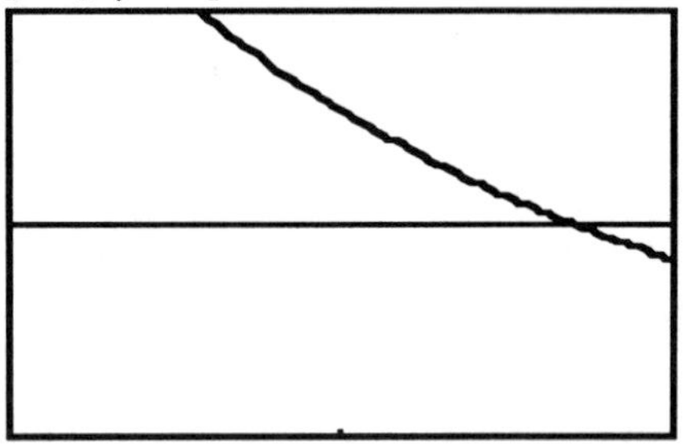

Zooming in on the intersection point we find t = 8451.28 years. So the fossilized organism lived approximately 8451 years ago.

8.1 Exercises

Draw the graphs of the following functions. For each function find the y-intercepts.

1. $f(x) = 7^x$.
2. $g(x) = (1.25)^x$.
3. $h(x) = e^{3x}$.
4. $f(x) = 4^{x/4}$.
5. $k(x) = 2^x - \left(\frac{1}{2}\right)^x$.
6. $g(x) = 3^{-x^2}$.

Using the functions $f(x) = 3^x$ and $g(x) = \left(\frac{2}{3}\right)^x$ determine the graphs of each of the following.

7. $h(x) = 2f(x)$.
8. $k(x) = -3f(x - 1)$.
9. $r(x) = f(2x + 1) - 4$.
10. $h(x) = \frac{1}{2}g(x)$.
11. $k(x) = -g(x + 2)$.
12. $r(x) = g(-x + 4) + 2$.

13. Draw the graphs of $f(x) = (2.5)^{2x+1} - 1$ and $g(x) = (2.5)(6.25)^x - 1$. Explain why there is no difference in the graphs.

14. The half-life of a radioactive substance is 125 days and initially we have 4.25 grams of the substance.
    (a) Find a function of the form $q(t) = q_0e^{-kt}$ that gives the amount of the substance at any time t.
    (b) Draw the graph of q(t) for $t \geq 0$.
    (c) How long will it take (in days) for 3.75 grams to decay?

15. Suppose a piece of a human skull is found and it is discovered to contain only 1% of the original amount of ^{14}C. When did the human die?

16. If an amount P is deposited in an account paying an annual interest rate r compounded m times a year, then the amount A at any time t (in years) is given by

$$A(t) = P\left(1 + \frac{r}{m}\right)^{mt}.$$

Assuming the annual interest rate is 8.5% determine the time required to double an initial amount $1000 if the interest is compounded
(a) Quarterly.
(b) Monthly.
(c) Daily.
(d) Hourly.
(e) Every minute.
(Hint: How many hours or minutes are there in a year?)

17. Consider the function $A(t) = Pe^{rt}$. If P = 1000 and r = .085 what value of t gives A(t) = 2000? Compare your answer to answers in 16 (a), (b), (c), (d), and (e). This is the formula for continuously compounded interest.

18. A bounded growth function or logistic function is a function of the form

$$f(t) = \frac{A}{B + Ce^{-kt}}$$

where k > 0. A logistic function that models the United States population is given by

$$f(t) = \frac{13.927793}{0.02494779 + 0.15832470e^{-0.18327249t}}$$

where t = 0 is the year 1900, t = 1 is the year 1910, etc.
(a) Compare the values of f(0), f(1), . . . , f(8) with the actual population given in the table in exercise 9 of section 3.3.
(b) Draw the graph of f(t) for $0 \le t \le 30$. During what years will the United States population grow fastest?
(c) According to this model what will the population of the United States be in the year 2000? What about in the year 2010? What about the year you turn 65?
(d) What value does f(t) approach as t becomes very large
(e) The value in part (d) is called the carrying capacity. Keeping in mind that the land area of the United States is 3,539,289 square miles, comment on the value of the carrying capacity.
(f) In what year will the population of the United States double its value in 1980? What percent of the carrying capacity is this? How old will you be in that year?

19. The thickness of a page of a typical textbook is approximately 0.004 inches. Tear a piece in half and stack the halves and tear them in half again. Stack the quarters and tear again in half, stack the eighths and tear again, etc.
(a) Write a function h(n), that gives the height of the stacks as a function of the number of times you tear the stacks in half.

(b) The average distance from the Earth to the Moon is 238857 miles. How many times do you have to tear and stack the halves so that the stack reaches to the Moon?

8.2 Logarithmic Functions

(McKeague: Sections 10.1, 10.2, 10.3)

Simply put, a logarithmic function is the inverse of an exponential function. This implies the following definition:

$$\log_a v = u \text{ means } a^u = v.$$

Stated differently, u, the logarithm base a of v, is the power a must be raised to produce v. On almost all scientific calculators (and the graphics calculators are no exception) there are two logarithmic functions:

Natural logarithmic function: $\ln v = u$ **means** $e^u = v$

Common logarithmic function: $\log v = u$ **means** $10^u = v$.

The figures below are the graphs of $y = \ln x$ and $y = \log x$, drawn with the default range setting. On the same graphs are $y = e^x$ and $y = 10^x$, showing that the logarithmic functions are inverses of the exponential functions.

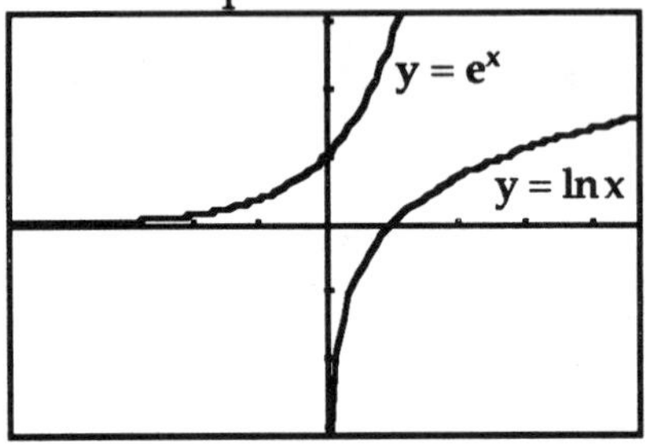

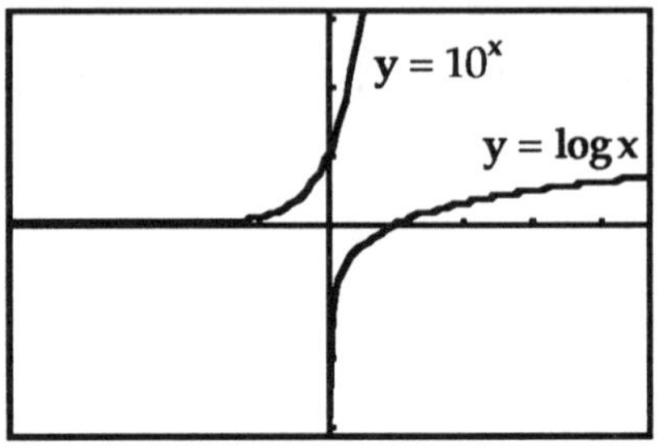

From these graphs it appears that either of these logarithmic functions has as its domain all positive real numbers. This is because there is no value of u such that e^u or 10^u is negative.. Also we see that the x-intercept of both functions is (1, 0). This is because $e^0 = 1$ so $\ln 1 = 0$.

Of all the logarithmic functions the natural logarithmic function enjoys the status of being the most widely used. It is this function that we will look at in detail.

- **Example 1:** Create a table of values for the natural logarithmic function and from this table determine what happens as x approaches zero and as x becomes very large.

x	lnx
.0000001	-16.1181
.000001	-13.8155
.00001	-11.5129
.0001	-9.2103
.001	-6.9078
.01	-4.6052
.1	-2.3026
1	0
10	2.3026
100	4.6052
1000	6.9078
10000	9.2103
100000	11.5129
1000000	13.8155

From the table it appears that as x approaches zero lnx approaches $-\infty$. In fact, $\ln 10^{-23} = -52.9595$. As x approaches infinity, lnx becomes large as well, but not very fast.

The fact that lnx grows slowly is evident from the above table. This can be shown more dramatically by using the fact that lnx is the inverse function of e^x. For example $\ln x = 100$ provided $e^{100} = x$. The value of e^{100} is 2.6881×10^{43}. So for lnx to be as large as 100, x must be as large as 2.7×10^{43}.

In the previous section in example 6.1.2 we found how long it takes for an initial investment to double given a certain annual interest rate. Let's see how the natural logarithmic functions can be used to solve this problem.

- **Example 2:** An amount P is deposited in a savings account paying an annual interest rate of r compounded m times per year. The amount in the account in any year t is given by the function $A(t) = P\left(1 + \frac{r}{m}\right)^{mt}$. Construct a function that gives the number of years required for the initial amount P to increase to qP, $q > 1$. For $r = 0.0825$ (8.25%) draw the graphs for $q = 2$ and $1 \le m < 365$, i.e., interest compounded yearly to daily. Is there a significant difference in the time required for P to double given the different compounding schemes?

The time t for P to increase to qP satisfies the equation $qP = P\left(1 + \frac{r}{m}\right)^{mt}$. Dividing out P and taking the natural logarithm of both sides produces $\ln q = mt\ln\left(1+\frac{r}{m}\right)$. Here we have used the fact that $\ln(v^k) = k\ln v$. Solving this equation for t gives

$$t = \frac{\ln q}{m\ln\left(1+\frac{r}{m}\right)}.$$

Setting q = 2 and r = 0.0825 we have

$$t = \frac{\ln 2}{m\ln\left(1+\frac{0.0825}{m}\right)}.$$

Here the independent variable is m and the dependent variable is t. Set the range to be [1, 365] x [8, 9] with **Xscl** = 30 and **Yscl** = .2 and draw the graph of $y = \frac{\ln 2}{x\ln\left(1+\frac{0.0825}{x}\right)}$. This gives the graph

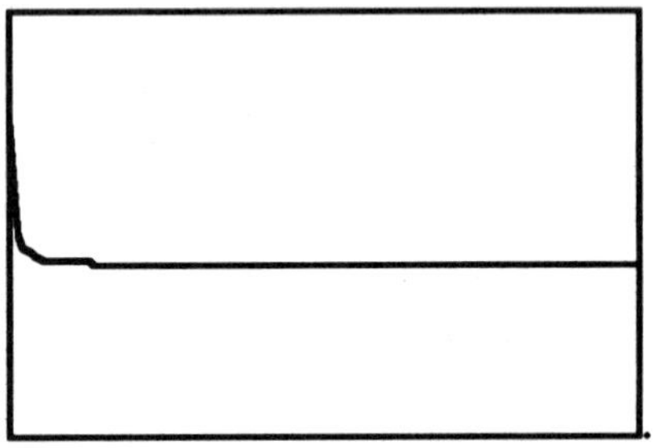

This graph tells us the doubling time of the investment as a function of the number of times per year the interest is compounded. Note that for x = 1, compounding yearly, it takes y = 8.744 years to double and for x = 12, compounding monthly, it takes y = 8.431 years to double. By using the trace, x = 12 is very near to the point where the graph begins to flatten out. We can conclude that the doubling time does not significantly change when compounding monthly or daily. For x = 365, the doubling time is 8.403 years and for x = 12 y = 8.431 years, a difference of 0.028 years or about 10 days.

The natural logarithm is also used to compare the relative sizes of two parts x, y of a single system, (for example, comparing the weight of an organism to its body volume or the mass of a planet to its escape velocity). In many situations the relationship between x and y is given by the formula $y = cx^m$ for some values of c and m. In many cases y and x have different dimensions and hence such a formula is called an allometric formula. If we take the logarithm of both sides of this equation we have $\ln y = m\ln x + \ln c$. Here we have used the fact $\ln(uv) = \ln u + \ln v$ and $\ln(v^k) = k\ln v$. If we denote $\ln y$ by Y, $\ln x$ by X, and $\ln c$ by b this equation now reads Y = mX + b, i.e., Y is a linear function of X with slope m and y-intercept b.

- <u>Example 3:</u> The planets of the solar system orbit the sun in an elliptical path with the sun at one focus. See the figure. The table below gives the orbital period, T, and length, d, of the semi-major axis of the orbit for each of the 9 planets in our solar system and Halley's comet.

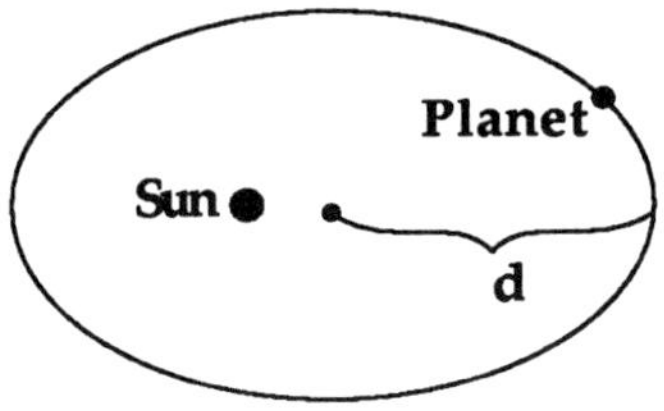

Planet	d (millions of km)	T (years)
Mercury	57.9	0.241
Venus	108	0.615
Earth	149	1
Mars	228	1.881
Jupiter	778	11.861
Saturn	1430	29.457
Halley's Comet	2690	75.733
Uranus	2870	84.008
Neptune	4500	164.784
Pluto	5810	248.35

Determine a formula that relates T and d.

If we assume that $T = cd^m$ then $\ln T = m\ln d + b$ where $b = \ln c$. If such a relationship holds, then if we plot the ordered pairs $(\ln d, \ln T)$ they should nearly lie on a straight line. Using the range setting [4, 9] x [-2, 6] and plotting these pairs give the graph

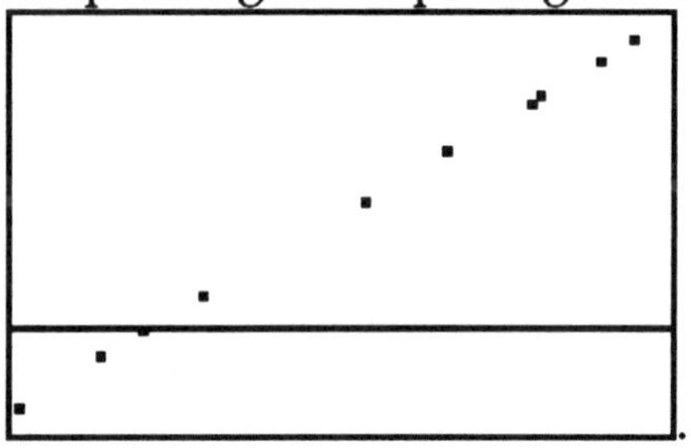

These points appear to lie on a straight line. To find the slope m we evaluate

$$m = \frac{\ln(248.35) - \ln(.241)}{\ln(5910) - \ln(57.9)} = 1.4998.$$

So $\ln T = 1.4998\ln d + b$. To find b we solve $\ln(248.35) = 1.4998\ln(5910) + b$ giving $b = -7.5100$. Therefore $\ln c = -7.5100$ or $c = e^{-7.5100} = 0.00055$. Thus $\ln T = 1.4998\ln d - 7.5100$ and if we exponentiate both sides we have $T = 0.00055d^{1.4998}$. If the graph of the equation $y = 1.4998x - 7.5100$ over the above graph we see that there is an almost perfect fit.

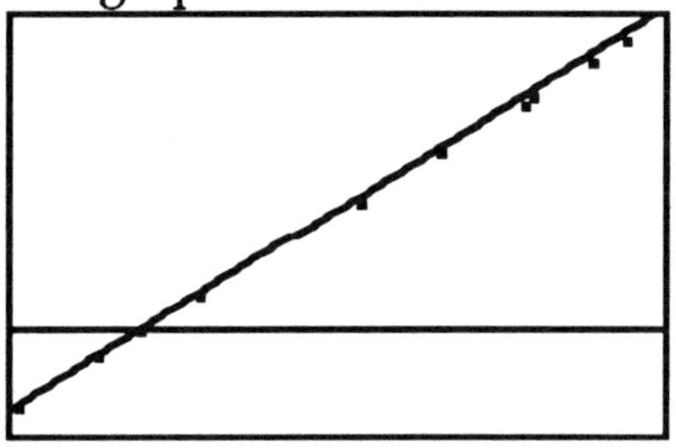

If we set the range to be [50, 6000} x [0, 260] and draw the graph of $T = 0.00055d^{1.4998}$ along with the original data the figure at the top of next page is produced.

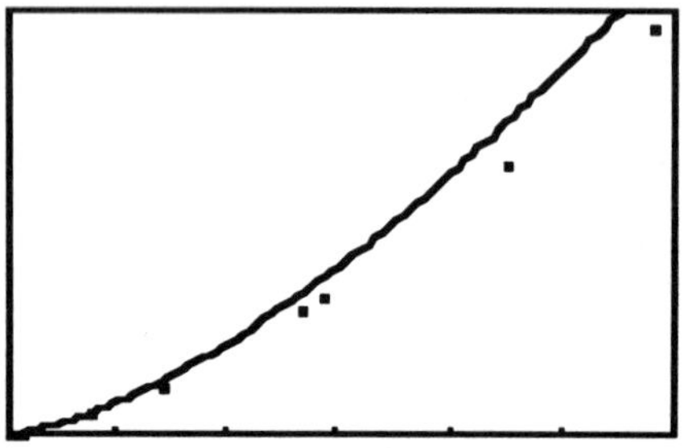

Johannes Kepler in the year 1618 guessed that $T = cd^{3/2}$.

Each of the graphics calculators has the natural logarithmic function, base e, and common logarithmic function, base 10, but how do we draw graphs and obtain values of logarithmic functions with bases different from e and 10? The answer to this lies in the change of base formulas:

$$\log_b v = \frac{\ln v}{\ln b} \quad \text{or} \quad \log_b v = \frac{\log v}{\log b}.$$

So to draw the graph of $y = \log_2 x$ simply draw the graph of $y = \dfrac{\ln x}{\ln 2}$. The following graph show the graphs of $y = \ln x$, $y = \log_2 x$, $y = \log_7 x$ and $y = \log_{1/2} x$. Notice that for the base 1/2 the function is decreasing.

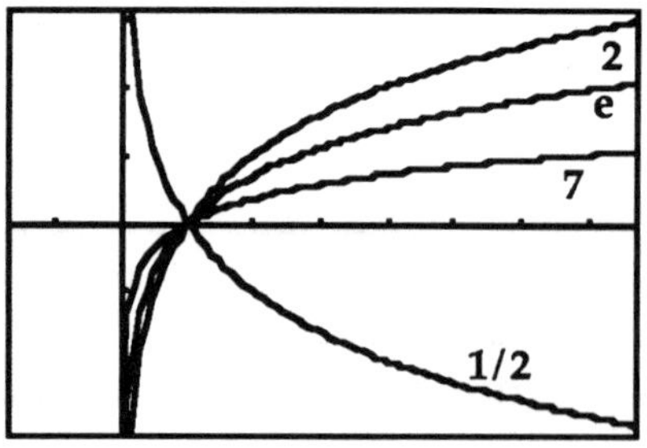

8.2 Exercises

Draw the graph of each of the following functions.

1. $f(x) = 2\ln(x + 1)$.
2. $g(x) = \ln(2x + 1)$.
3. $h(x) = \log(1 - x)$.
4. $k(x) = \log(|x| + 1)$.
5. $f(x) = \ln\left(\dfrac{x+1}{x-1}\right)$.
6. $g(x) = \log\left(\sqrt{x^2 - 1}\right)$.
7. $k(x) = \log_5\left(4 - x^2\right)$.
8. $h(x) = 4\log_2\left(\dfrac{x}{x+1}\right)$.

9. The following table gives masses, radii, and escape velocities of the 9 planets in our solar system and our moon. The escape velocity is the speed with which an object must travel in order to escape the gravitational pull of the planet.

Planet	Radius (miles)	Mass (tons)	Escape Velocity (mph)
Moon	1079.90	8.112×10^{20}	5291.42
Mercury	1511.61	3.507×10^{21}	9310.00
Mars	2100.19	7.079×10^{21}	11225.51
Pluto	3422.20	1.191×10^{22}	11208.01
Venus	3760.23	5.683×10^{22}	23221.23
Earth	3958.13	6.595×10^{22}	25006.71
Neptune	14110.75	1.133×10^{24}	54702.18
Uranus	14580.58	9.577×10^{23}	49565.80
Saturn	36165.47	6.270×10^{24}	80596.63
Jupiter	41520.83	2.097×10^{25}	134411.01

(a) Plot the pairs (Mass, Escape Velocity). Does it appear that Escape Velocity, V, is related to mass, M, by a formula of the form $V = cM^m$? Suppose this is true; is $0 < m < 1$ or is $m > 1$? Explain your answer.

(b) Plot the pairs (lnM, lnV). Do these points fall nearly on a straight line? As in example 6.2.3 find the slope and intercept of this line and draw the graph of this line on the same graph as the points (lnM, lnV). How well does the line fit the points?

(c) Using the slope and intercept in part (b) determine c and m so that $V = cM^m$. Draw the graph of $V = cM^m$ ($y = cx^m$) on the same graph as the pairs (M, V). How well does it fit these points?

(d) From the equation $V = cM^m$ determine the escape velocity of Titan, a moon of Saturn, whose radius is 1477.97 miles and mass is 1.312×10^{21}. Compare your value to the actual escape velocity of 5774.05 mph.

(e) Repeat steps (a) - (d) for the radius, R, and escape velocity, V. Which variable, mass or radius, predicts escape velocity better? Why?

10. (a) For the population data in exercise 9, of Chapter 5, section 2, plot the pairs (x, lnP), where x takes the values 0, 1, 2, 3, . . . , 8, and P are the populations values. Do these points fall nearly on a straight line? If so, then $\ln P = mx + b$. Assuming this is true, show that this implies that $P = ce^{mx}$, where $c = e^b$.

(b) As in example 6.2.3, find a slope and intercept for the line and plot the pairs (x, lnP) with this line and also pairs (x, P) along with the function $P = ce^{mx}$.

(c) When will the population of the United States in 1980 double according to the model $P = e^{mx}$?. Compare this with the bounded growth model in exercise 18 of section 8.1.

Investigating Exponential Growth

How much should I save for retirement? When should I start saving? How important is the interest rate? Use the formula for the future amount of an annuity to investigate these questions. Consider a tax-deferred fund that is idealized because it includes neither service charges nor withdrawals.

The amount of an annuity A_n is given as

$$A_n = P\frac{(1+i)^n - 1}{i}$$

where n is the number of regular equal payments of size P, and i is the interest rate per time period. If the interest rate is 12% per year and the regular payment is made monthly, then $i = 0.12/12 = 0.01$.

Investigate the amount of an annuity with interest rates of 3%, 6% and 12% after x years, if you make monthly payments of $100. Use your graphs to complete Table 1 and help you with the investigation. The equation for the amount of the annuity at 3%, $A_n = 100\dfrac{\left(1+\frac{0.03}{12}\right)^{12x} - 1}{\frac{0.03}{12}}$, *is shown on the equation screen (Figure 1). Figure 2 is the graph of the amount of annuity, y, and the number of years, x. The cursor indicates the value of the annuity at 20 years.*

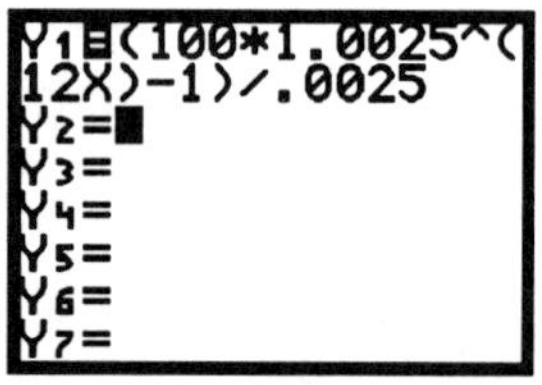

Figure 1

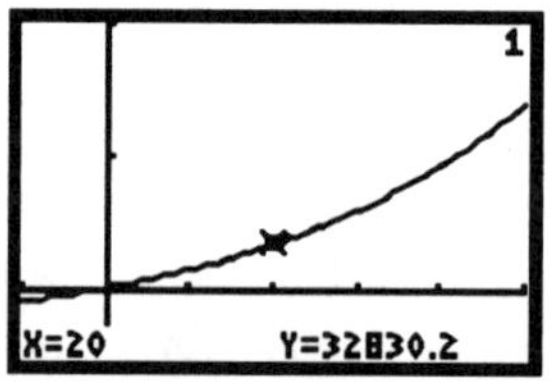

Figure 2

Time (years)	P (monthly payments)	Total Invested in Annuity	A_n given	annual	interest
			3 %	6 %	12 %
20	$100	$24,000	$30,830.20		
30	$100				
40	$100				

Table 1

Investigate the amount *of an* annuity with regular monthly payments of \$100, \$200, *and* \$400 and an annual interest rate of 6%. *Use your graphs to complete Table 2 and help you with the investigation.*

Time (years)	Interest (annual)	P monthly payments		
		\$100	\$200	\$300
20	6 %			
30	6 %			
40	6 %			

Table 2

Suppose Sandra decides to begin saving \$100 per month at age 25. However, her twin decides to spend money when she is young and begin to save when she is 50. How much will the twin need to save per month to have the same amount in an annuity as Sandra when they reach age 65? Assume the rates of the annuities are the same.

Discuss the amount of an annuity *that requires a single annual payment of \$1200 compared to the amount of an annuity at the same rate that has a monthly payment of \$100.*

Answers to Exercises

Section 1.2

1. 3.4	2. -3.333333333	3. -5
4. 110	5. 2.291025971	6. 44.19417382
7. 0.2891946824	8. 6	9. 5
10. 7.071067812	11. 1.1974474356	12. -1.414213562
13. 0.3333333333	14. -21.05623926	15. 2.703244951

Section 1.3

1. 0.4375, 0, 10.5
2. $\sqrt{13} = 3.605551275$
3. 0, 0, -24, 0
4. 9
5. 0.3580868716, 0.3539964407, 0.3535975958, 0.3535092075 0.3531125503, 0.3492411225

Section 1.4

1. (a) V = 150.7964474, S = 175.9291886 (b) V = 7.853981634, S = 21.99114858
2. Continution the process given in the hint, the 'halving' idea the following table is produced.

H	R	V
5	5	392.6990817
5	10	1570.796327
5	7.5	883.5729338
5	8.75	1202.640938
5	8.125	1036.971013
5	7.8125	958.7379924
5	7.96875	997.4100730
5	8.046875	1017.125136
5	8.0078125	1007.274103
5	7.98828125	1002.366563
5	7.978515625	999.9172872
5	7.983398438	1001.141551
5	7.980957030	1000.529325

3. Add the lines

CASIO

```
"Q":V÷(RS)→Q◢
```

TI-81

```
Disp "Q"
V/(RS)→Q
Disp Q
```

R = H, V/(RS) = 1/4
R = H/2, V/(RS) = 1/3
R = 4H, V/(RS) = 1/10
R = H/3, V/(RS) = 3/8

Note: $\dfrac{V}{RS} = \dfrac{\pi R^2 H}{R[2\pi R(R + H)]} = \dfrac{H}{2(R + H)}$.

4. Programs

CASIO

```
"L"?→L:"W"?→W
"H"?→H
"VOL"
LWH→V ◢
"AREA"
2(LW+LH+WH)→S ◢
```

TI-81

```
Disp "L"
Input L
Disp "W"
Input W
Disp "H"
Input H
"VOL"
LWH→V
Disp V
"AREA"
2(LW+LH+WH)→S
Disp S
```

(a) V = 70
S = 118

(b) V = 32.4
S = 84.48

5. Add the lines

CASIO

```
"Q":V÷(LS)→Q ◢
```

TI-81

```
Disp "Q"
V/(LS)→Q
Disp Q
```

If L/H = W/H = c, i.e., L = cH, W = cH, then

$$\frac{V}{LS} = \frac{LWH}{2L(LW + LH + WH)} = \frac{c^2H^3}{2cH(c^2H^2 + cH^2 + cH^2)}$$

$$= \frac{c^2H^3}{2c^2H^3\,(c + 2)} = \frac{1}{2(c + 2)}.$$

Section 1.5

1. Range: [-3, 3] x [-4, 9]

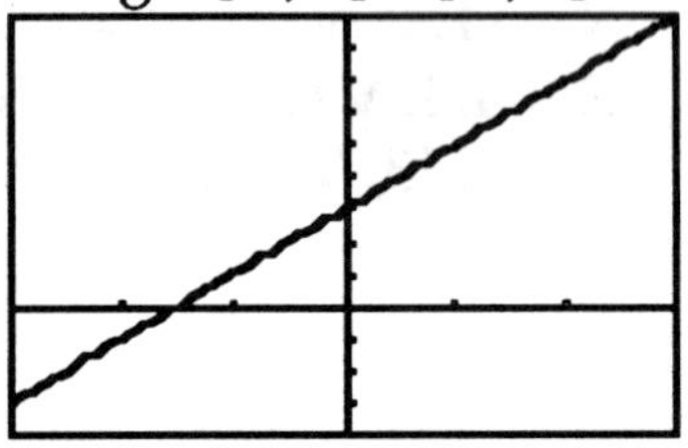

2. Range: [-2, 4] x [-20, 5]

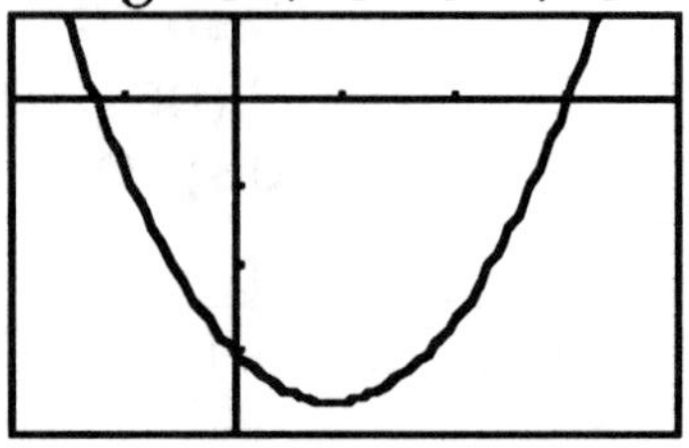

3. Range: [-3, 3] x [-2, 5].

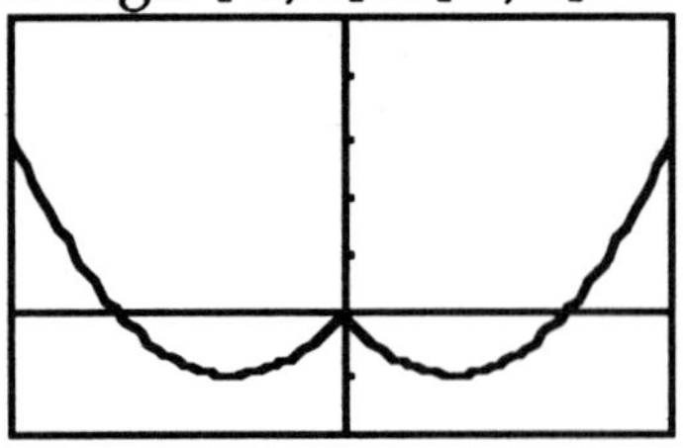

4. Range: [-4.7, 4.7] x [-3.1, 3.1]

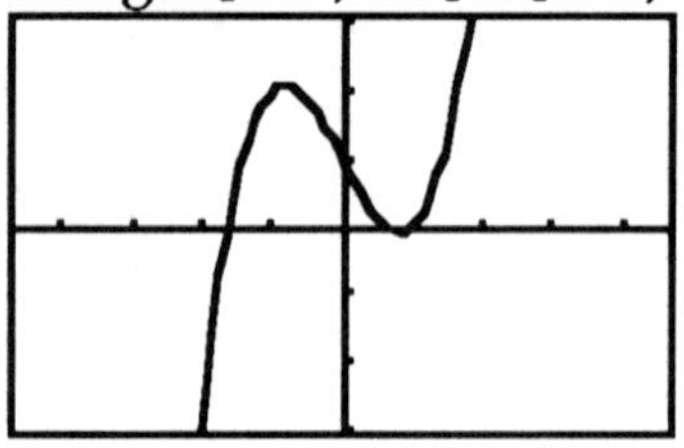

5. $x = -1, 4$.
6. $x=-0.3909838$, $x = 2.5576505$
7. $x = \pm\sqrt{5}$
8. Equation is $1000 = \pi R^2\, 5$, so graph $y = 5\pi x^2 - 1000$ and find the positive x-intercept.

Answers to Investigating Order of Operations

1A. 4 1B. -19

Group numerator and denominator by enclosing each in parentheses.

2A. 2 2B. 2 2C. 18 2D. 18

Refer to page 4. Rule (Order of Operations)
When evaluating a mathematical expression, we will perform the operations in the following order, beginning with the expression in the innermost parentheses or brackets and working our way out.

1. Simplify all numbers with exponents, working from left to right if more than one of these expressions is present.
2. Then, do all implicit multiplications.
3. Then, do all multiplications and divisions left to right.
4. Perform all additions and subtractions left to right.

Section 2.1

1. [-14.1, 14.1] x [-9.3, 9.3]

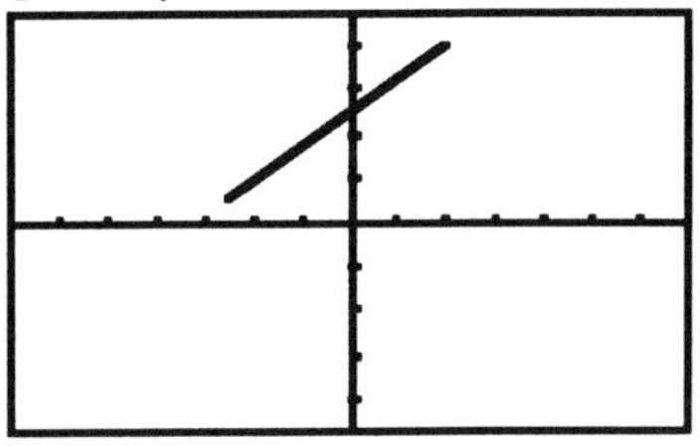

2. [-14.1, 14.1] x [-9.3, 9.3]

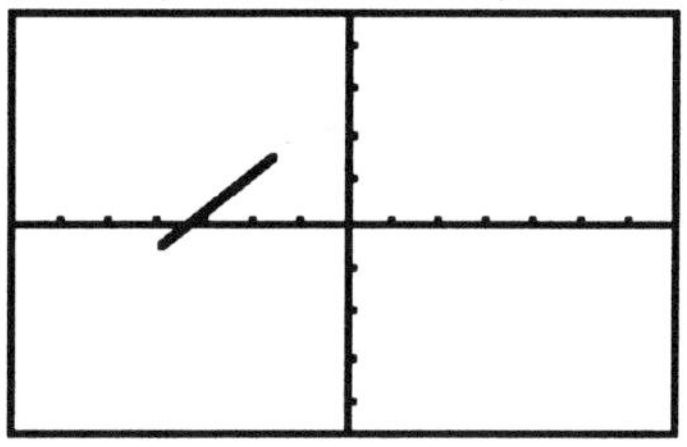

3. [-2.35, 2.35] x [-1.55, 1.55]

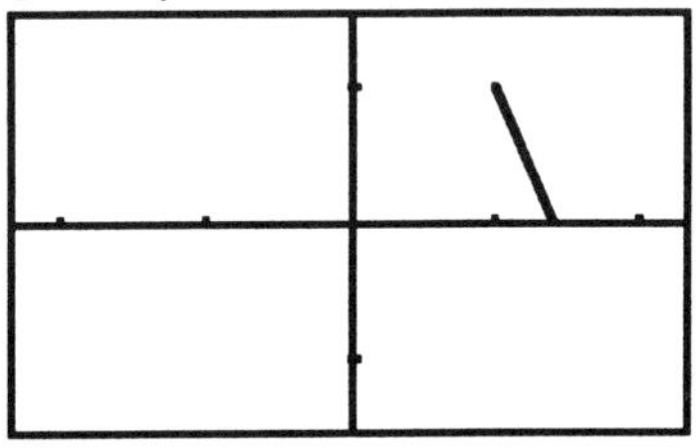

4. [-2.35, 2.35] x [-1.55, 1.55]

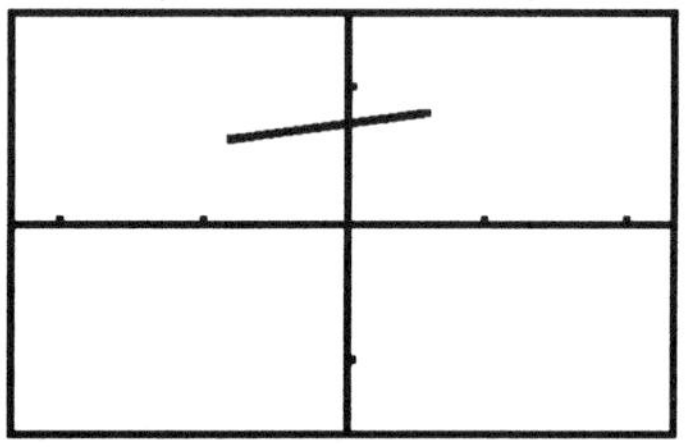

5. Range: [-.5, 1.5] x [-3, 4.5]

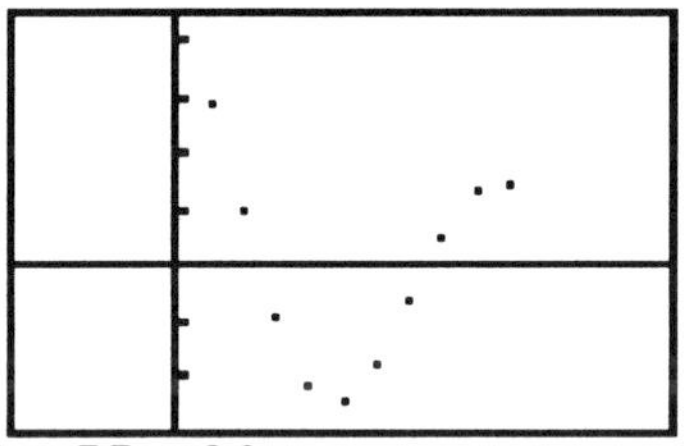

6. $c = 1, 1 \leq H \leq 20$
$1 \leq R \leq 20.$

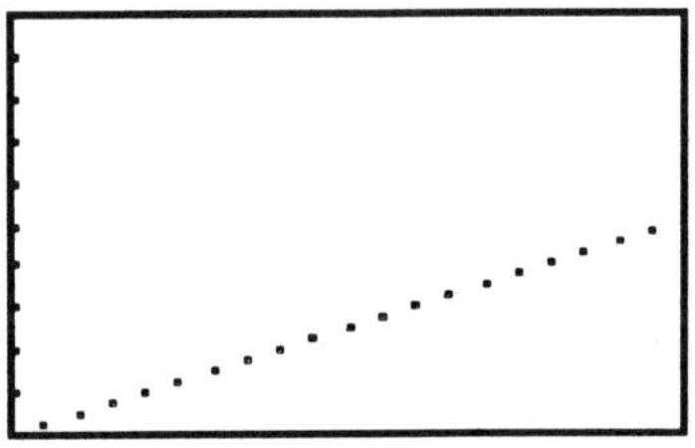

[0, 20] x [0, 10]

$c = 2, 1 \leq H \leq 20$
$1 \leq R \leq 40$

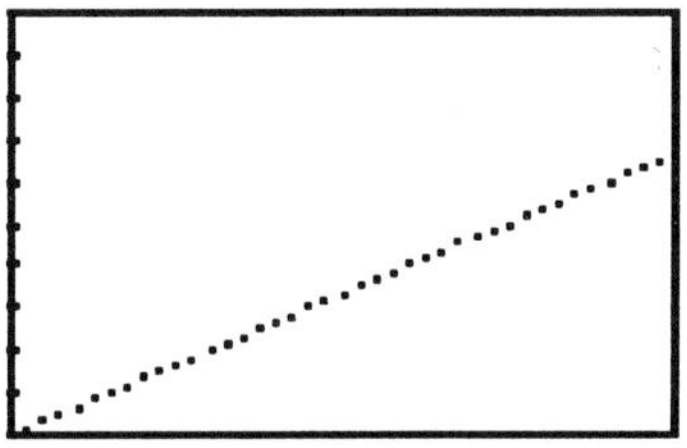

[0, 40] x [0, 10]

$c = 1/2, 1 \leq H \leq 20$
$1 \leq R \leq 10$

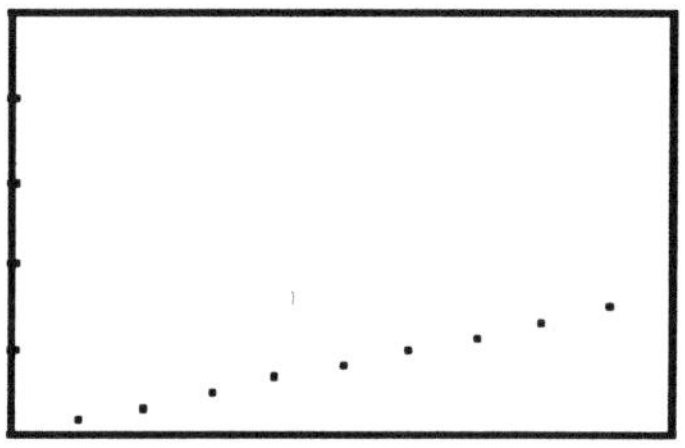

[0, 10] x [0, 10]

$c = \pi, 1 \leq H \leq 20$

$1 \leq R \leq 20\pi$

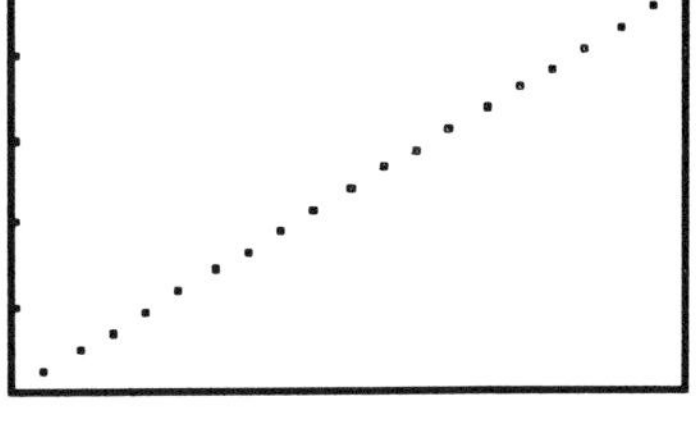

Range: [0,20π] x [0, 8].

It appears R and V/S are related by a linear function.
In fact, $\frac{V}{S} = \left(\frac{c}{2 + 2c}\right) R.$

7. [-5, 5] x [-12, 12].
8. [-5, 5] x [-2, 1]
9. [-5, 5] x [-1, 6].
10. [-5, 5] x [-20, 5].

11. (Payment No., Interest)

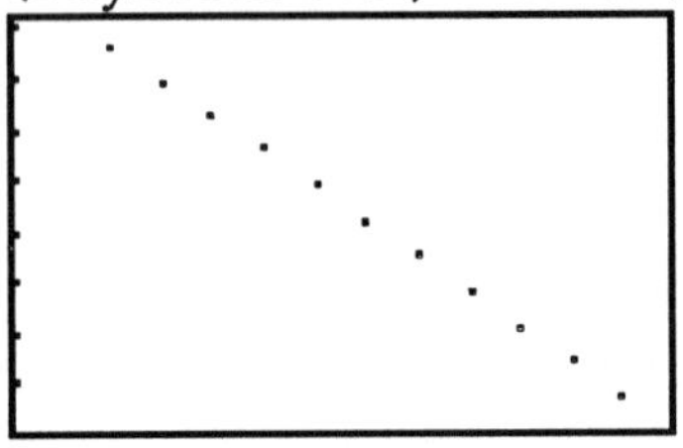

[0. 13] x [0, 41.25]

(Payment No., Reduction)

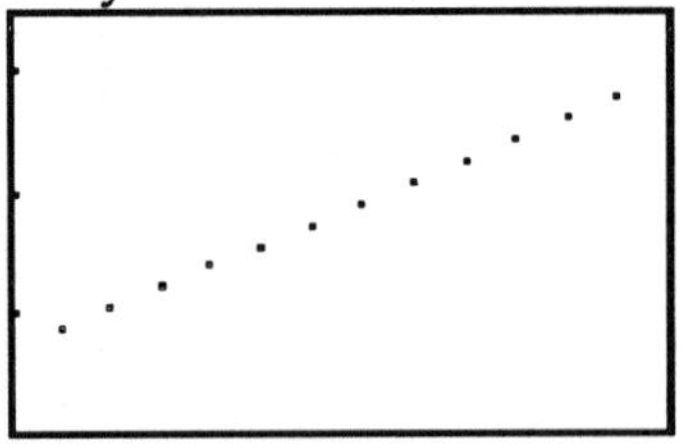

[0, 13] x [380, 450]

(Payment No., Balance)

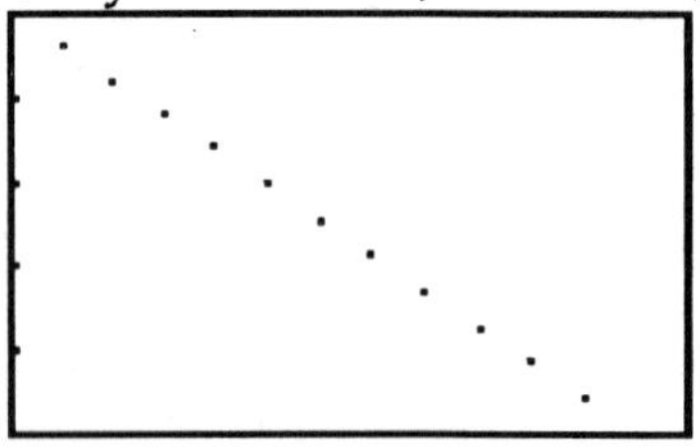

[0, 13] x [0, 5000]

It appears that these graphs are those of linear functions.

Section 2.2

For these exercises, the range setting given is such that the solution can be seen.

1. [-5, 5] x [-2, 8], x = 1.
2. [5, 15] x [5, 8], x = 0.
3. [-20, -1] x [6, 18], x = -9.8.
4. [-4.7, 4.7] x [3, 9], x = -0.2.
5. [-2, 0] x [-10, -3], x = -1.
6. [-4.7, 4.7] x [-3.1, 3.1], x = -0.8.

Section 2.3

For these exercises, the range setting given is such that the solution set can be seen.

1. [-5, 4] x [-6, 1], x > -1.25.
2. [-2, 8] x [7, 13], x ≥ 3.
3. [-6, 4] x [-5, 1], x < -2/13.
4. [-4.7, 4.7] x [-3.1, 3.1], -67/30 < x < -1.1.
5. [-4.7, 4.7] x [-3.1, 3.1], x ≥ 2 or x ≤ 2/3.
6. [-4.7, 4.7] x [-.1, 6.1], 0 < x ≤ 4/3 or 2 ≤ x < 10/3.

Answers to Investigating Absolute Value

a	b	c	d	Number of Solutions
2	0	0	2	2
2	2	0	2	1
2	4	0	2	0
2	2	0	4	2
2	4	0	4	1
2	6	0	4	0
2	1	2	2	1
2	4	1	2	infinite number
2	6	1	2	0
2	0	-1	0	0
2	-2	-1	0	infinite number
2	-4	-1	0	1
2	0	-2	0	1
2	0	-1/2	2	2
2	4	-1/4	-2	0

Table 1

There are 0,1,2, or an infinite number of solutions.
If $c = 0$
and $b > d$, then there are no solutions
$b = d$, then there is one solution
$b < d$, then there are two solutions.
If $c = \pm 1$, then there are 0, 1, or infinite number of solutions.
If $c \neq 0, \pm 1$, then there are 0, 1, or 2 solutions.

Section 3.1

1. $(x + 7)(x - 4)$
2. $4(x + 3)(x - 2)$.
3. $(2x + 1)(x + 2)(x - 2)$.
4. $(x + 3)(x - 2)(x - 3)$.
5. $(x + 3)(x - 2)(x - 5)$.
6. $2(2x + 9)(x + 2)(x - 1)$.
7. $(2x + 1)(x - 1)(x - 3)(x - 12)$.
8. $(2x + 3)(x + 10)(3x - 1)(x - 4)$.

Section 3.2

1. 5
2. -1.4, 9
3. 1, 6.

Section 3.3

1. Zeros in intervals: [-2, -1], [-1, 0], [1, 2]. Zeros are: -1.41421356, -.5, 1.41421356.

2. Zero in interval [6, 7]. Zero is 6.72180656.

3. Zeros in intervals [-2, -1.5], [-1, -.5], [.5, 1], [1.5, 2]. Zeros are ±1.84785906, ±0.7653669.

4. Zeros in intervals [-2, -1.5], [1.5, 2]. Zeros are -1.69028449, 1.50506841, and 1.

5. Zero in the interval: [-2.8, -2.7]. This zero is -2.75. Other zeross: 1.5, 1.4, -3.

6. Zeros in intervals [-1, -.9], [-.75, -.5], [-.5, 0], [0, .5], [.5, .75], [.9, 1]. Zeros are ±0.96592583, ±0.70711068, ±0.25881905.

7. Use the range setting: [-4, 6] x [-2000, 200]. Time taken by each calculator to draw the graph of p(x) using the power form and the nested form.

TI-81:	Power form:	14 seconds	CASIO:	Power form:	27 seconds
	Nested form:	8 seconds		Nested form:	8 seconds.
TI-82:	Power form:	9 seconds			
	Nested form:	6 seconds.			

8. These are the values interpolated by the polynomials, but because of round off error, you won't get these values exactly when the polynomials are evaluated.
$p_1(0) = 75.995, p_1(8) = 226.505.$
$p_2(0) = 75.995, p_2(4) = 131.669, p_2(8) = 226.505.$
$p_3(0) = 75.995, p_3(2) = 105.711, p_3(6) = 179.323, p_3(8) = 226.505.$
$p_5(0) = 75.995, p_5(1) = 91.972, p_5(3) = 123.203, p_5(5) = 150.697, p_5(7) = 203.212,$
$p_5(8) = 226.505.$
$p_6(0) = 75.995, p_6(1) = 91.972, p_6(2) = 105.711, p_6(3) = 123.203, p_6(5) = 150.697,$
$p_6(7) = 203.212, p_6(8) = 226.505.$
$p_8(0) = 75.995, p_8(1) = 91.972, p_8(2) = 105.711, p_8(3) = 123.203, p_8(4) = 131.669,$
$p_8(5) = 150.697, p_8(6) = 179.323, p_8(7) = 203.212, p_8(8) = 226.505.$

Population predictions for the years 1990, x = 9, and 2000, x = 10.

$p_1(9) = 245.319, p_1(10) = 246.133$	$p_6(9) = 143.658, p_6(10) = -272.731$
$p_2(9) = 256.333, p_1(10) = 288.609$	$p_8(9) = 93.388, p_8(10) = -1330.951$
$p_3(9) = 253.409, p_3(10) = 282.795$	
$p_5(9) = 217.148, p_3(10) = 129.238$	

Answers to Investigating Areas Given a Fixed Perimeter

P	$A = w\left(\frac{P}{2} - w\right)$	Maximum Area	Width	Length
30	$-w^2 + 15w$	56.25	7.5	7.5
20	$-w^2 + 10w$	25	5	5
40	$-w^2 + 20w$	100	10	10
50	$-w^2 + 25w$	156.25	12.5	12.5

Table 1

The area is a maximum when the rectangular area is a square or when the length = width. The length = width = $\frac{\text{perimeter}}{4}$.

P	$A = w\,(P - 2w)$	Maximum Area	Width	Length
30	$w(30 - 2w)$	112.5	7.5	15
20	$w(20 - 2w)$	50	5	10
40	$w(40 - 2w)$	200	10	20
50	$w(50 - 2w)$	312.5	12.5	25

Table 2

The maximum area is a rectangle whose length is twice the width. The width = $\frac{\text{perimeter}}{4}$ and the length = $\frac{\text{perimeter}}{2}$.

Answers to Investigating Rational Exponents and Roots

If the base is positive, place rational exponents in parentheses. For example $64^{2/3}$ would be entered into the calculator as 64^(2÷3). to get the correct answers.

The answers to 2A, 2B, and 2D are 4 which is correct. The answer to 2C is approximately -1.26 which is the $\sqrt[9]{-8}$. If the base is negative, separate the exponent into its numerator and one over the denominator as in 2A and 2B. Do not use the x^2 after the rational exponent.

The domain of f(x) is $x \geq 0$. The rational exponent with a number greater than 1 in the numerator is not defined over negative numbers. The domain of g(x) is all real numbers and g(x) best describes $y = x^{2/3}$.

Answers to Investigating the Domain of Rational Expressions

The calculation $\frac{0}{5}$ is equal to 0, but $\frac{5}{0}$ will give a calculator answer of error or a complex number depending on your calculator. The calculation $\frac{5}{0}$ is not defined for real numbers.

Rational Expression	Domain
$\frac{1}{3x + 12}$	x = {Real numbers, x ≠ -4}
$\frac{1}{x^2 + 5x + 6}$	x = {Real numbers, x ≠ -3, -2}
$\frac{x - 2}{2x - 5}$	x = {Real numbers, x ≠ 5/2}

Table 1

The domain will be undefined where there are vertical asymptotes.

Rational Expression	Actual Domain	Apparent Domain from Graph
$\frac{x-1}{x^2+3x-4}$	x = {Real numbers, x ≠ -4, 1}	x = {Real numbers, x ≠ -4}
$\frac{2x-5}{-2x^2+15x-25}$	x = {Real numbers, x ≠ 5, 5/2}	x = {Real numbers, x ≠ 5}
$\frac{x}{5x^3-125x}$	x = {Real numbers, x ≠ 5, ±5}	x = {Real numbers, x ≠ ±5}

Table 2

If the numerator is a factor of the denominator, then the value of x where that factor equals zero will not appear as a vertical asymptote on the graph. So, at this point the graph will appear to be defined even though closer investigation will show it has a "hole" there. No, you could not determine the domain solely from the graph because you could not decide if common factor(s) in the numerator and denominator had been reduced.

Section 4.1

1. $y = (7x - 38)/3$
[-23.5, 23.5] x [-15.15, 15.5]

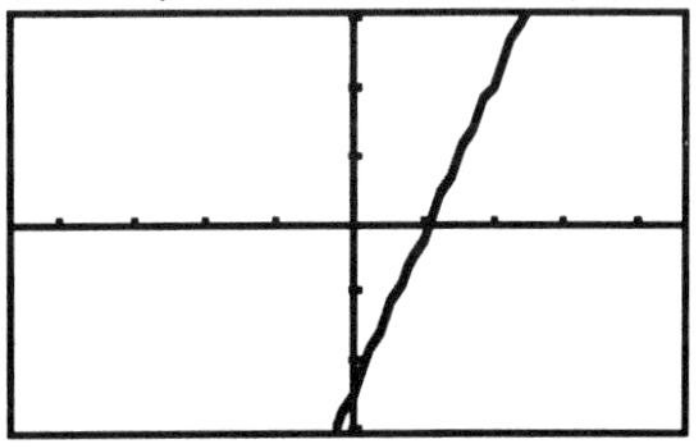

2. $y = -(51x+239)/32$
[-23.5, 23.5] x [-15.5, 15.5]

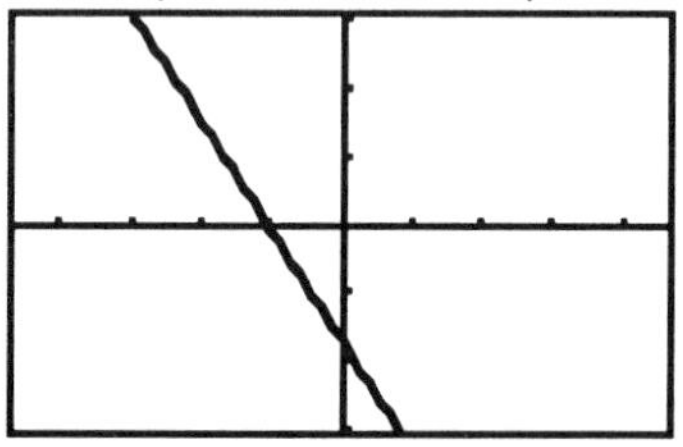

3. $y = -4x + 7$
[-23.5, 23.5] x [-15.15, 15.5]

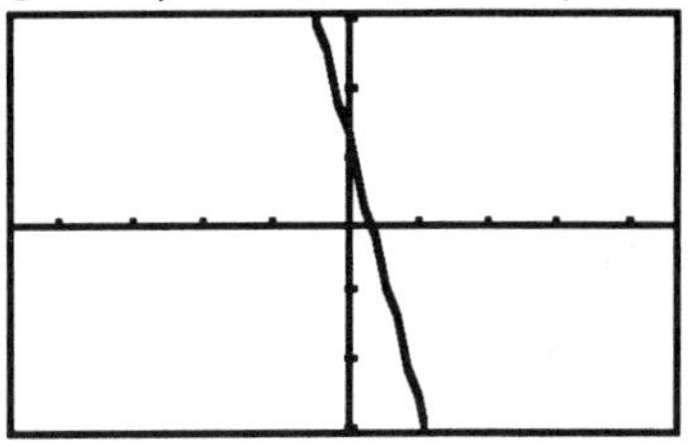

4. $y = 8x/5 - 11$
[-23.5, 23.5] x [-15.5, 15.5]

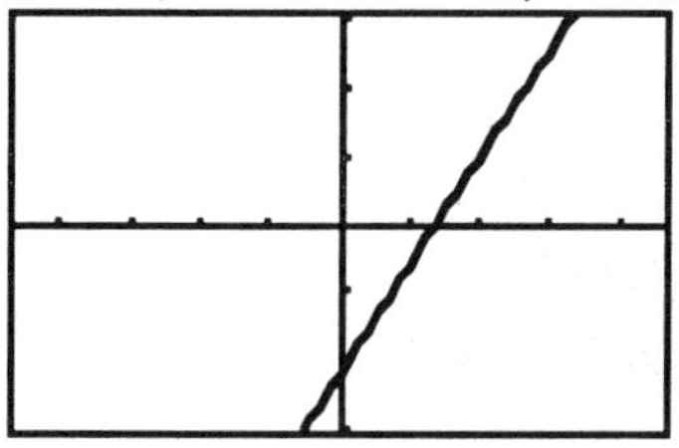

5. Slope of given line: m = -2/3,
Given line, l_1: y = -(2x + 4)/3
Parallel line, l_2: y = (-2x + 11)/3

Range: [-9.4, 9.4] x [-6.2, 6.2]

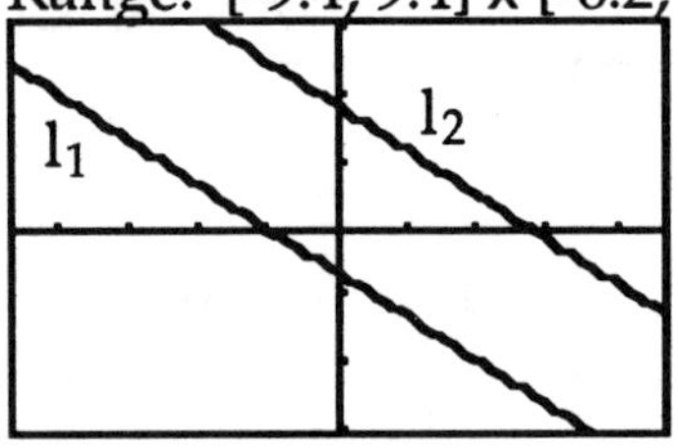

6. Slope of given line: m = 4/5
Given line l_1: y = (4x + 12)/5
Perpedicular line, l_2: y = -5x/4 – 1

Range: [-9.4, 9.4] x [-6.2, 6.2]

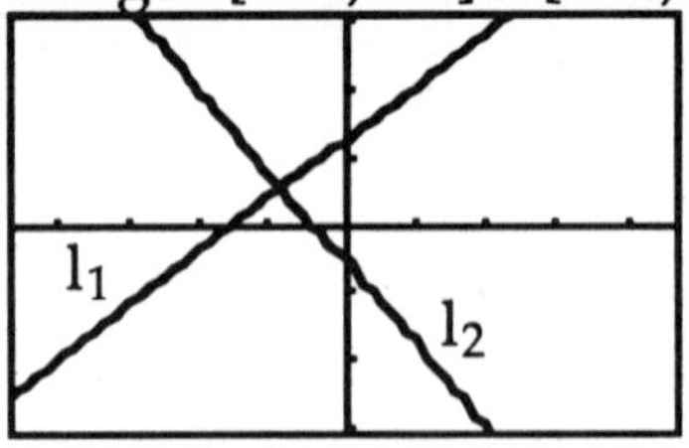

7. Let 1, 2, 3, . . . , 12, be the months January, February, March, . . . , December.

$$\text{Price: } p(x) = \begin{cases} -5.3125x + 90.3125 & \text{if } 1 \le x \le 5 \\ 3.0357x + 48.5714 & \text{if } 5 \le x \le 12 \end{cases}.$$

Range: [1, 12] x [60, 90].

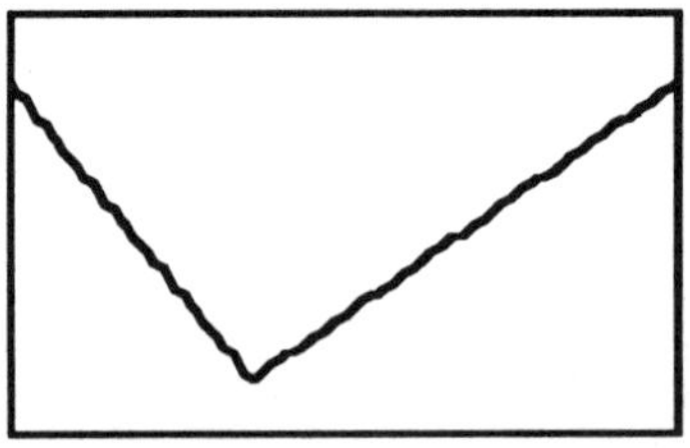

Section 4.2

1. All (x, y) below the graph of
y = (2x – 1)/3.
Range: [-4.7, 4.7] x [-3.1, 3.1]

TI-81/82: **Shade(Ymin,(2X–1)/3,1)**
fx-7700G: **Graph Y<(2X–1)÷3**

2. All (x, y) below the graph of
y = -2x + 7/2.
Range: [-4.7, 4.7] x [-3.1, 3.1]

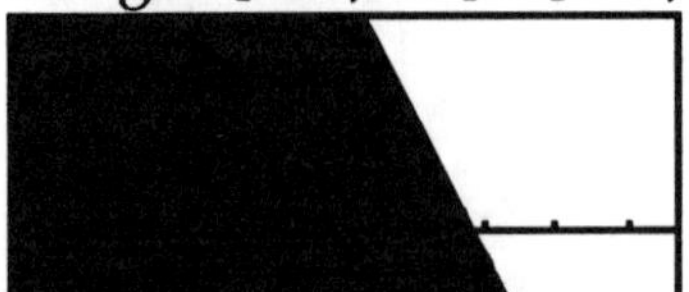

TI-81/82: **Shade(Ymin,-2X+7/2,1)**
fx-7700G: **Graph Y≤-2X+7÷2**

3. All (x, y) below the graph of y = x + 2, above the the graph of y = 2x – 1, and above the graph of y = -x/2 – 4.
Range [-9.4, 9.4] x [-6.2, 6.2]
On the TI-81/82, shade the opposite of the solution. Commands are

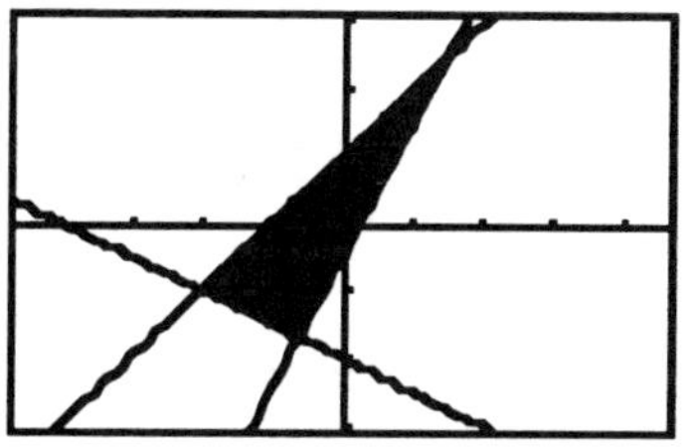

Shade(X+2, Ymax,2)
Shade(Ymin,2X–1,3)
Shade(Ymin,-X/2–4,4)
fx-7700G: **Graph Y≤X+2:Graph Y≥2X–1:Graph -X÷2–4**
This is the graph shown to the right.

4. All (x, y) above the graph of y = 2x – 1, below the graph of y = (4 – x)/3 and to the right of x = -3.
Range: [-11.75, 11.75] x [-7.75, 7.75]
TI-81/82: **Shade(2X–1,(4–X)/3,1,-3,Xmax)**
fx-7700G: **Graph Y≥2X–1:Graph Y<(4–X)÷3:Graph Y≤1E8(x+3)**

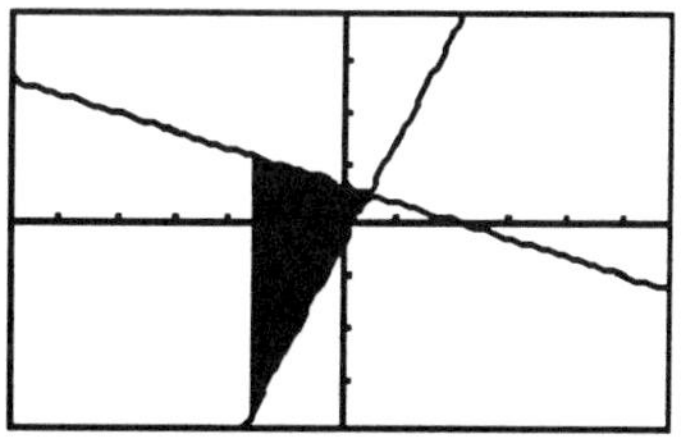

5. All (x, y) below the graph of y = 6 – x, above the graph of y = (x – 12)/6, and below the graph of y = (12 + 3x)/2.
Range: [-9.4, 9.4] x [-6.2, 6.2]
On the TI-81/82, shade the opposite of the solution. Commands are:

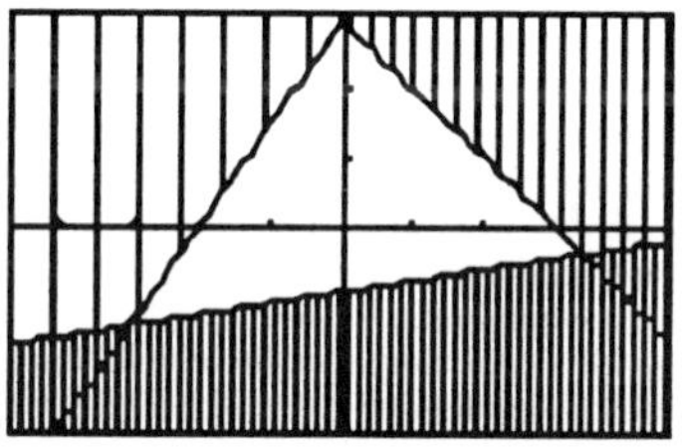

Shade(Ymin,(X–12)/6,2)
Shade(6–X,Ymax,3)
Shade((12+3X)/2,Ymax,4)
This is the graph shown to the right.
fx-7700G: **Graph Y≤6–X:Graph Y≥(X–12)÷6:Graph Y≤(12+3X)÷2**

Answers to Investigating Slope

Absolute Value Function	Point at Which Slope Changes	Slope to Left of Point	Slope to Right of Point	Equation That Graphs Slope of Absolute Value Function
$f(x) = \lvert x+5 \rvert$	(-5, 0)	-1	1	$f(x) = \dfrac{\lvert x+5 \rvert}{x+5}$
$f(x) = \lvert 3x+6 \rvert$	(-2, 0)	-2	2	$f(x) = \dfrac{\lvert 3x+6 \rvert}{x+2}$
$f(x) = \lvert 2x-5 \rvert$	(2.5, 0)	-3	3	$f(x) = \dfrac{\lvert 2x-5 \rvert}{x-\frac{5}{2}}$

Table 1

The equation that will graph the slope of f(x) is $m(x) = \dfrac{\lvert ax+b \rvert}{x+\frac{b}{a}}$.

Section 5.1

1. -3, -3/2.
2. -5, 4/3.
3. 1/5, -7/2.
4. $2 \pm \sqrt{3}$.
5. $(1 \pm \sqrt{5})/2$.
6. $1/3 \pm \sqrt{2}$.
7. $3 \pm 2i$.
8. $(2 \pm \sqrt{3}i)/5$.
9. $(1 \pm i)/\sqrt{2}$,

10. Set $u = x^2$, then $3u^2 + 16u - 35 = 0$. So $u_1 = 5/3$, hence $x_1 = \pm\sqrt{5/3}$.
Also $u_2 = -7$, hence $x_2 = \pm\sqrt{7}\,i$.

11. Set $u = \sqrt{x}$, then $u^2 - 4u - 3 = 0$. So $u_1 = 4.645751311$, $x_1 = u_1^2 = 21.58300524$.
Also $u_2 = -0.645751311$, but $x_2 = u_2^2$ is not a solution.

12. Set $u = 3^x$, then $3u^2 - 82u + 27 = 0$. So $u_1 = 27$, hence $3^{x_1} = 27$ or $x_1 = 3$.
Also $u_2 = 1/3$, hence $3^{x_2} = 1/3$ or $x_2 = -1$.

13. (a) $R = 1000/5\pi - 7.978845608$ (b) $R = 2.208024459$
(c) $\pi HR^2 - V = 0$, gives $R = \sqrt{V/\pi H}$.
$2\pi R^2 + 2\pi HR - S = 0$ gives $R = \left(\sqrt{H^2\pi^2 + 10\pi} - \pi H\right)/2\pi$.

14. $h = -16t^2 + 97t + 4.5$. Hits ground when h = 0, or t = 6.108542083 seconds.
Range: [0, 6.5] x [0, 155].

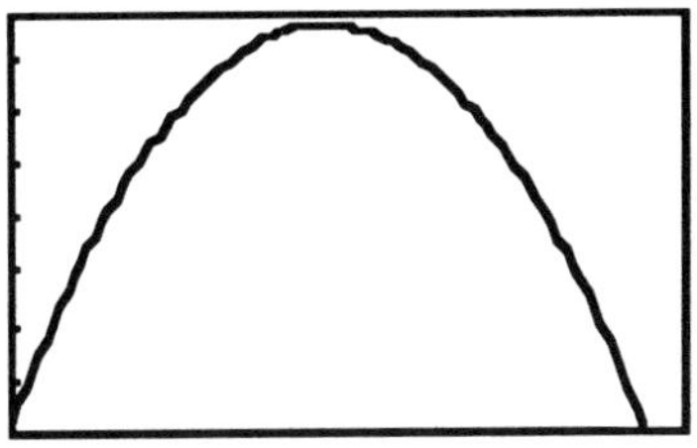

Tracing gives maximum height of 151.5. Zooming in with XFactor of 5 and YFactor of 20 gives t = 3.0315, h = 151.516.

15. (a) M = 0.9367. Using quadratic formula, horizontal distance travelled is 855.954435 feet. Range: [0, 860] x [0, 350].

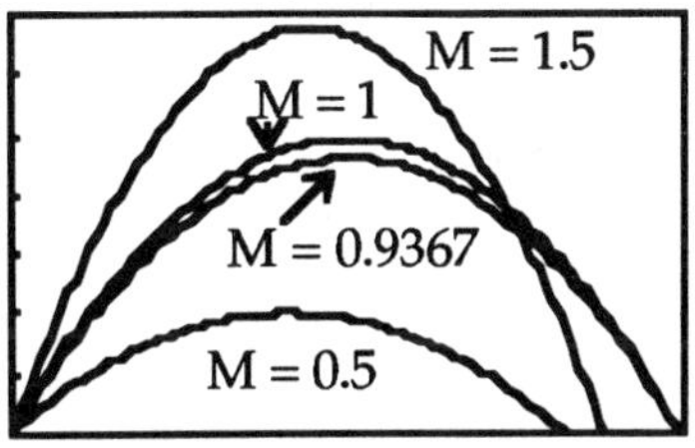

(b) W = 0, A = -0.003400999, B = 1, C = 5. Distance = 298.9489994.

(c) M = 0.45756

(d) Programs

CASIO

```
"M"?→M:"V"?→V
5→H:22→W
-16(1+M²)÷(V−W
   √(1+M²))²→A
VM÷(V−W√(1+M²))→B
H→C
Graph Y=AX²+BX+C ◢
Prog_
```

TI-81

```
All-Off
Disp "M"
Input M
Disp "V"
5→H
22→W
-16(1+M²)/(V−W
   √(1+M²))²→A
VM/(V−W√(1+M²))→B
H→C
DrawF "AX²+BX+C"
Pause
Prgm_
```

Set the range to be [0, 1300] x [0, 450]. The programs Prog_(CASIO) and Prgm_(TI-81/82) are quadratic formula programs that are used to find the distance.

Section 5.2

The range setting given is such that the solutions set can be viewed.

1. $[-5, 5] \times [-20, 5]$; $-5/3 < x < 3$.
2. $[-5, 8] \times [-5, 15]$; $-2 \le x \le 5$.
3. $[-1, 3] \times [3, 6]$; $3/5 \le x \le 1$, $x \ge 2.228285686$, $x \le -0.6282856857$.

4. $S = 2\pi R(R+H)$, $H = 5$. Solve $2\pi R^2 + 10\pi R \ge 150$. $R \ge 2.988464399$.
5. Solve $-16t^2 + 125t + 5 \ge 150$. $1.417015366 \le t \le 6.395484634$.
6. $2.372 < m < 2.582$ or $0.251 < m < 0.286$.

Section 5.3

The range setting is that given by the programs 2.1.1C or 2.1.1T, centered on the vertex and the value of **S** is given

1. Vertex: $(-2, -3)$, $[-6.7, 2.7] \times [-6.1, .1]$, **S** = 0.1, $y = 2(x + 2)^2 - 3$.
2. Vertex: $(5/2, 9/4)$, $[-2.2, 7.2] \times [-0.85, 5.35]$, **S** = 0.1 ,$y = -(x - 5/2)^2 + 9/4$.
3. Vertex: $(-4, 3)$, $[-8.7, 0.7] \times [-6.1, .1]$, **S** = 0.1 ,$y = -(x + 4)^2 - 3$.
4. Vertex: $(-6, 32)$, $[-10.7, -1.3] \times [28.9, 35.1]$, **S** = 0.1 ,$y = -(2/3)(x + 6)^2 + 32$.

5. Vertex: $(-1, -1)$, $[-3.7, 5.73] \times [-4.1, 2.1]$, **S** = 0.1.
6. Vertex: $(-1, 2)$, $[-5.7, 3.7] \times [-1.1, 5.1]$, **S** = 0.1.
7. Vertex: $(3/2, 1/4)$, $[-0.85, 3.85] \times [-1.3, 1.8]$, **S** = 0.5.
8. Vertex: $(-3/4, 3/2)$, $[-5.45, 3.95] \times [-1.6, 4.6]$, **S** = 0.1.
9 Vertex: $(-2, -4)$, $[-6.7, 2.7] \times [-7.1, -0.9]$, **S** = 0.1.
10. Vertex: $(-19/4, 1/2)$, $[-14.15, 4.65] \times [-5.7, 6.7]$, **S** = 0.2.
11 Vertex: $(9, 1/2)$, $[4.3, 13.7] \times [-2.6, 3.6]$, **S** = 0.1.
12. Vertex: $(3/2, 5)$, $[-7.9, 10.9] \times [-1.2, 11.2]$, **S** = 0.2.

13. (a) 145.4912348 feet (b) 145.4912348 feet
(c) 145.4912348 feet (d) y-coordinate of vertex is

$$c - b^2/4a = h - \left(\frac{v\,m}{v - w\sqrt{1 + m^2}}\right)^2 \frac{(v - w\sqrt{1 + m^2})^2}{-64(1 + m^2)}$$

$$= h + \frac{v^2m^2}{64(1 + m^2)}.$$

So maximum height depends only on the slope, initial velocity, and height. Not on the wind velocity.

(e) $w = .6(121) = 72.6$ feet/second. Horizontal distance traveled is 17.83344509 feet.

14. Range: [-10, 10] x [-8, 8]. Shape is the same, but different vertices.

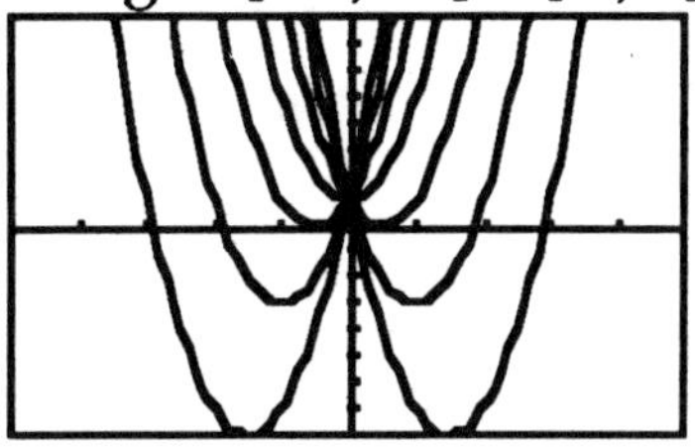

b = -6, -4, -2, 0, 2, 4, 6.

Range: [-10, 10] x [-8, 8]. Vertices appear to lie on a parabola.

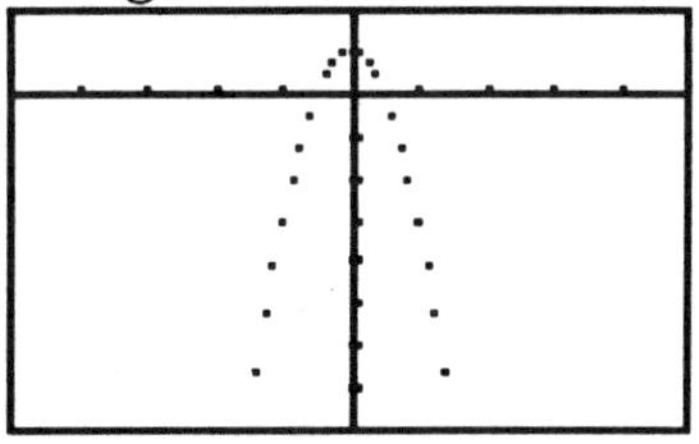

b = -6, -5.5, -5, -4.5, -4, . . . ,5.5, 6.

For the equation $y = x^2 + bx + 1$, the x-coordinate of vertex is $-b/2$ and y-coordinate is $1 - b^2/4$. If $x = -b/2$, then $b = -2x$ so $y = 1 - b^2/4 = 1 - (4x^2)/4 = 1 - x^2$. The equation of the curve where the vertices lie is $y = 1 - x^2$.

Answers to Investigating Quadratic Inequalities

Graph each side of the inequality as a separate function and find the x-coordinate of the points of intersection. Determine the range of x values that satisfy the conditions of the inequality.

1. $-2 \le x \le 5$

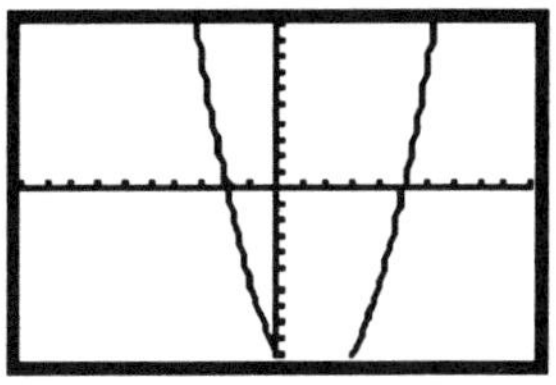

2. $x < -3.5$ or $x > 1$

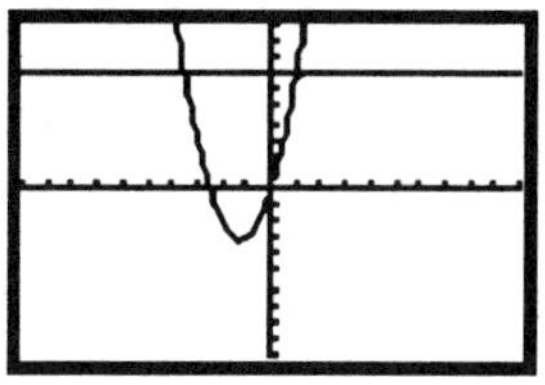

3. $x \le -2$ or $-1 \le x \le 3$

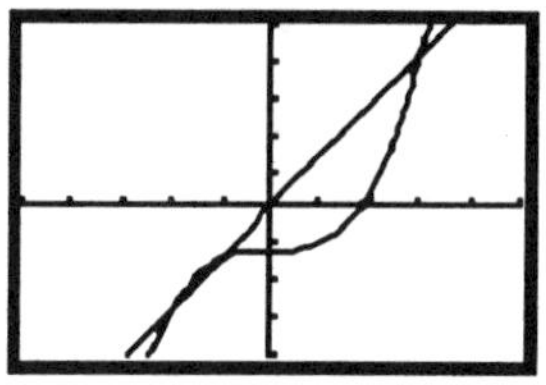

Section 6.1

1. $x = 11/9, y = 5/9$.
2. $x = 7/11, y = 61/22$.
3. $x = -8/21, y = 13/21$.
4. $x = 17/13, y = 7/13$.

Answers to Solving Systems of Linear Equations Using Matrices

Student research

1. x = 3, y = 1
2. x = 2, y = -1, z = 3
3. Although this is an inconsistent system with no solution this answer cannot be found directly from the calculator using a matrix equation. For this system the calculator will produce an error message which depends on the calculator model used.

This method is useful unless the linear system is dependent or inconsistent.

Section 7.1

1. a = -2, c=1, d = -3/2. 2. a=1/2, c = 2, d = 1. 3. a = 1/2, b = 1, c = -1/2, d = 1.

4. Range: [-9.4, 9.4] x [-6.2, 6.2] Range: [-9.4, 9.4] x [-6.2, 6.2] Range: [-9.4, 9.4] x [-6.2, 6.2]

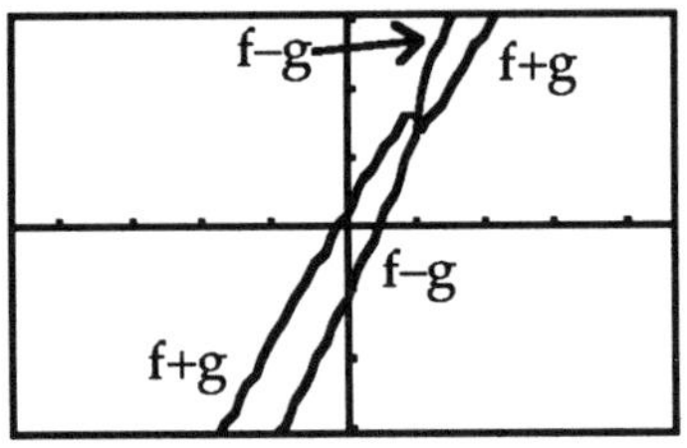

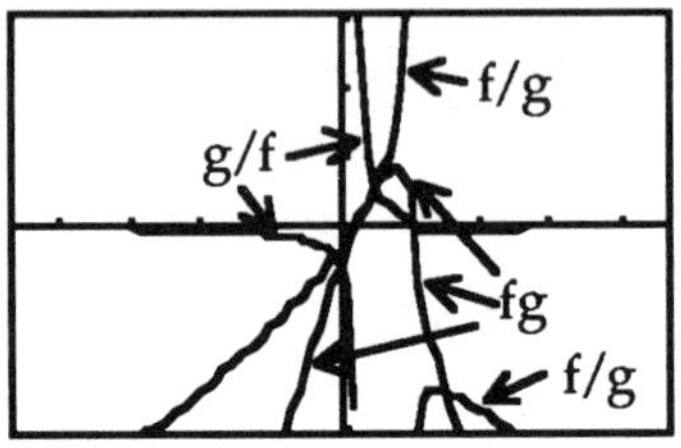

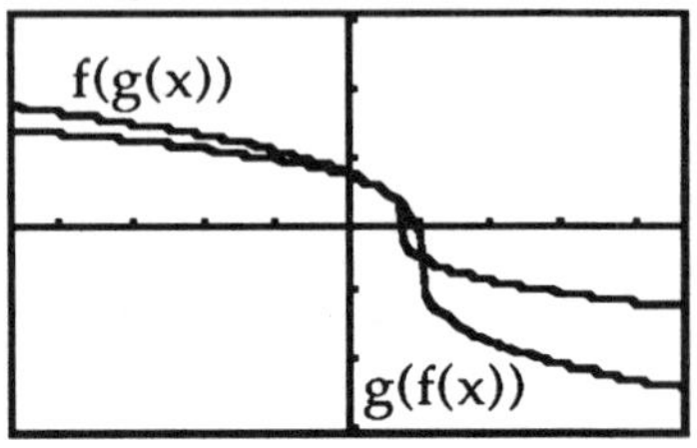

5. Range: [-9.4, 9.4] x [-6.2, 6.2] Range: [-9.4, 9.4] x [-6.2, 6.2] Range: [-9.4, 9.4] x [-6.2, 6.2]

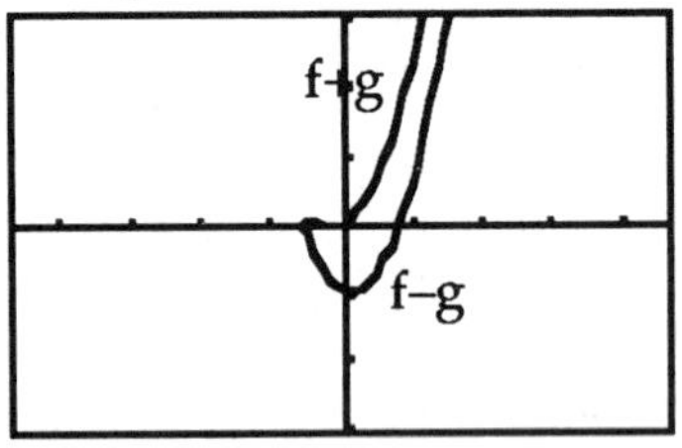

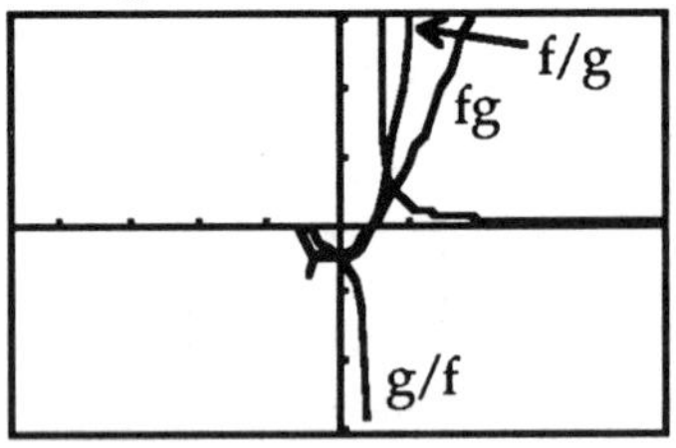

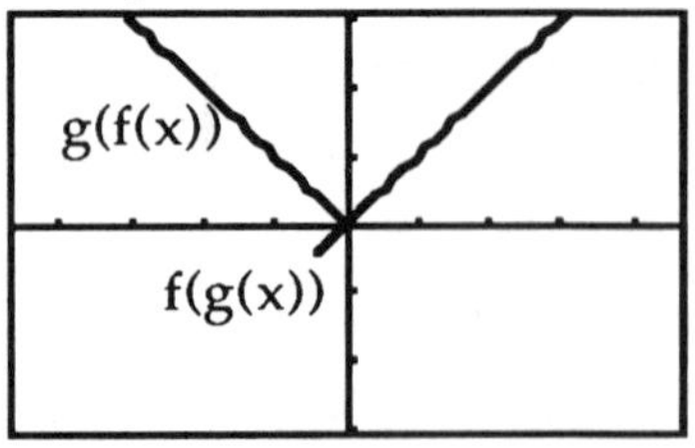

6. Range: [-24, 24] x [-16, 16] Range: [-31, 40] x [-14, 35] Range: Default

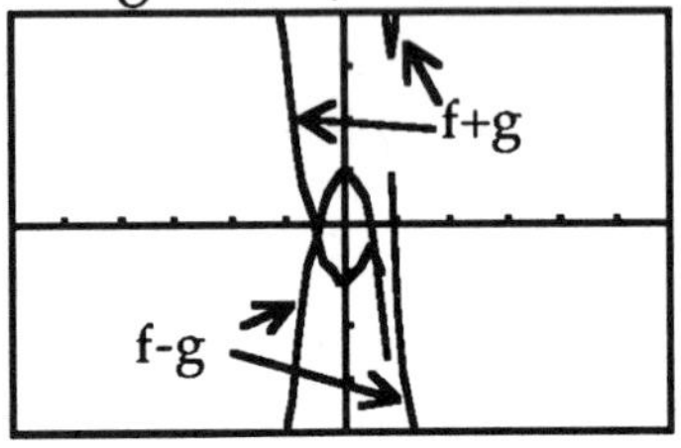

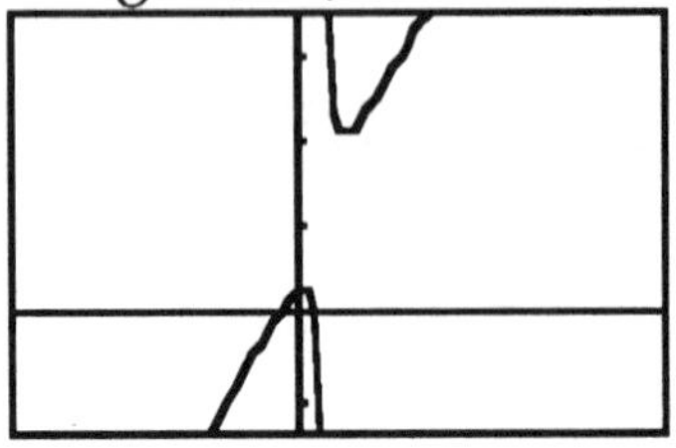

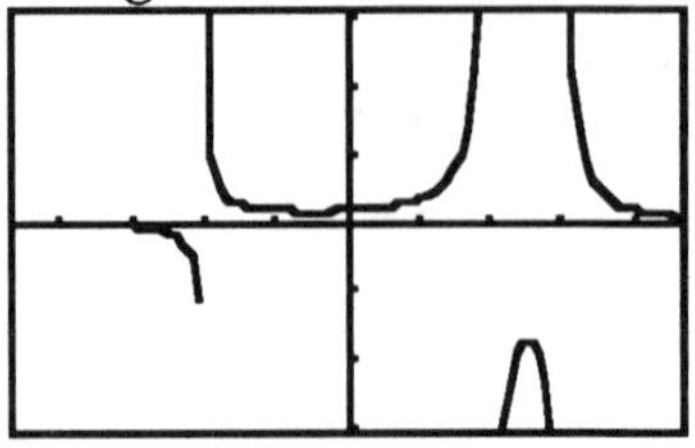

Range: [-9.4, 9.4] x [-6.2, 6.2]

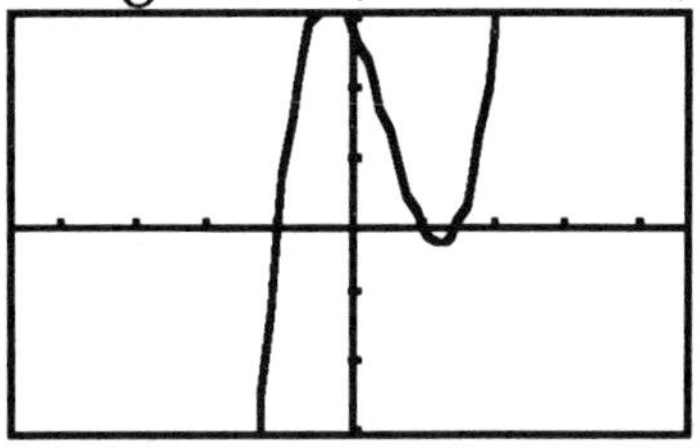

Range: [-9.4, 9.4] x [-6.2, 6.2]

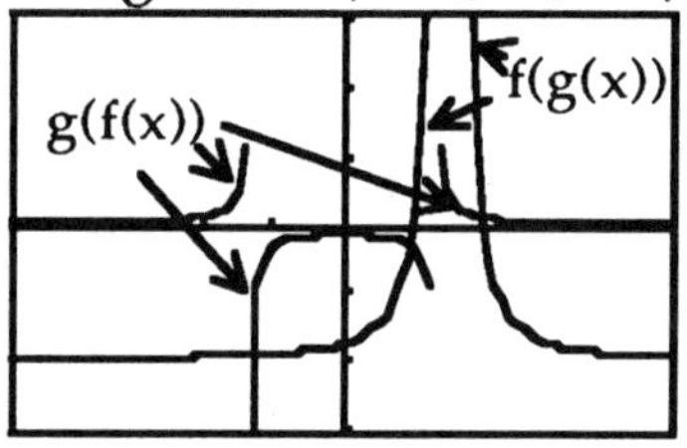

7. Range: [-9.4, 9.4] x [-6.2, 6.2].

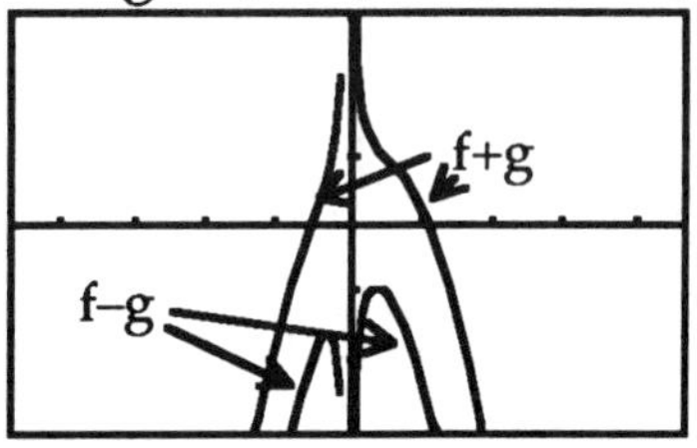

Range: [-9.4, 9.4] x [-6.2, 6.2]

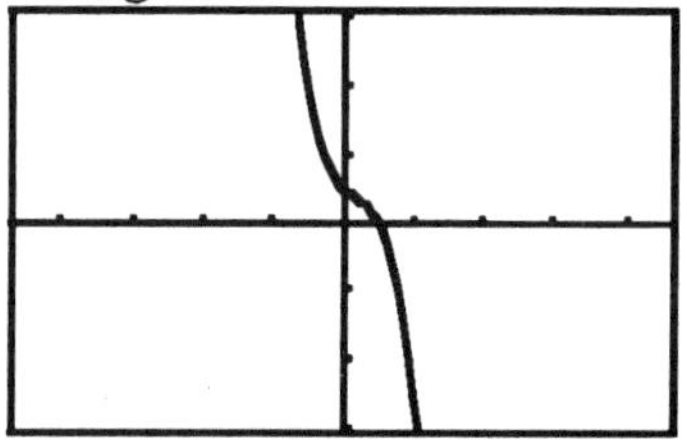

Range: Default

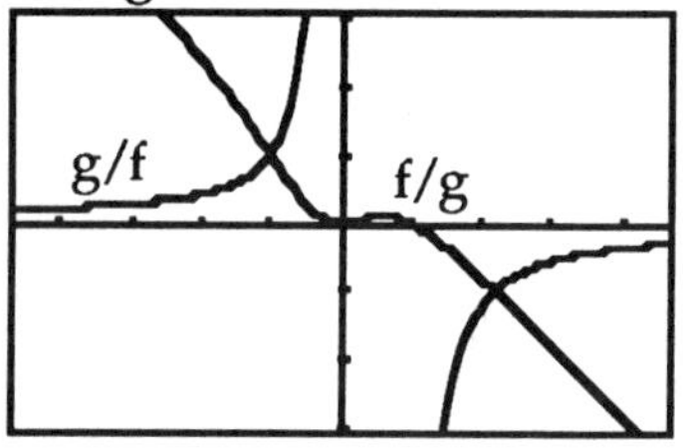

Range: [-8, 8] x [-10, 0]

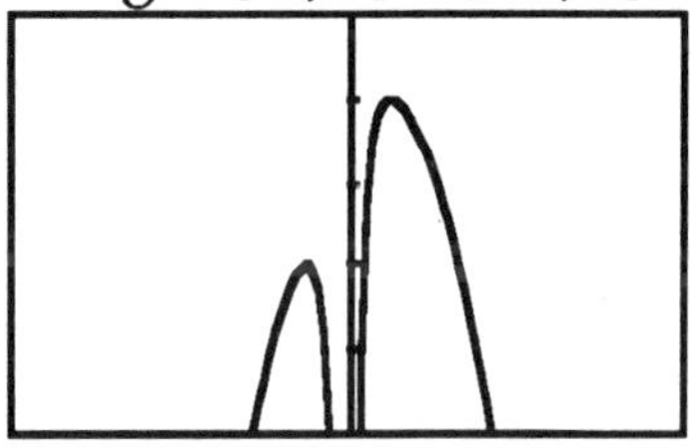

Range: [-5, 5] x [-10, 10]

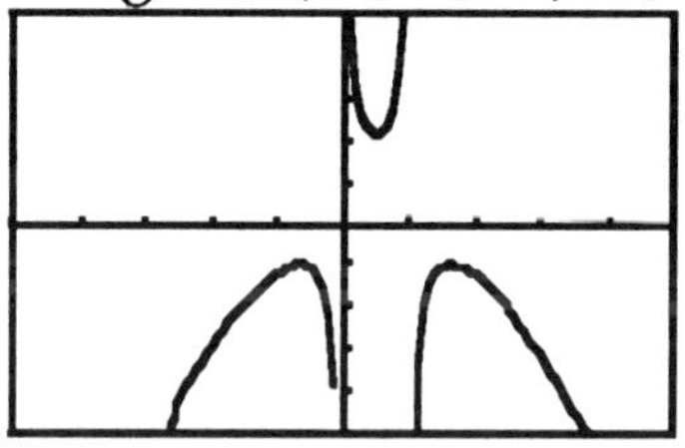

8. $\frac{V}{S} = \frac{cH}{2(c+1)}, \frac{S}{V} = \frac{2(c+1)}{cH}, \frac{V}{RS} = \frac{1}{2(c+1)}$.

Section 7.2

1. $V = 25h^3/27$, $S = 7.1168h^2$.
From the graph of S versus h, we see that $S = ah^2$ is a good guess. If $h = 1.5$, $S = 16.0128$, hence $a = 16.0128/(1.5)^2$. From the graph of V versus h, we see that $V = ah^3$ is a good guess. If $h = 1.5$, $V = 3.125$, hence $a = 3.125/(1.5)^3 = 25/27$.

2. $A = \pi ab = \pi a^2/2$. Try $P = ca^r$. If $a = 1$, $P = 4.8442$, so $c = 4.8442$. If $a = 2$, then solve $9.6884 = 4.8442(2^r)$ giving $r = 1$. So $P = 4.8442a$, thus $a = P/4.8442$. Therefore $A = \pi P^2/2(4.8442)^2$, i.e., $A = 0.6694P^2$.

3. (a) Plot of data. Range: [0, 12] x [75, 210].

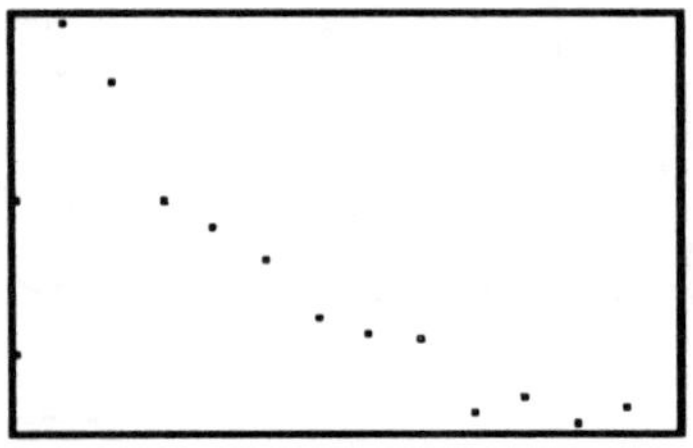

As in example 4.1.1, try plotting month (x) versus 1/Barrels per day (1/y).

Range: [0, 12] x [0, .013].

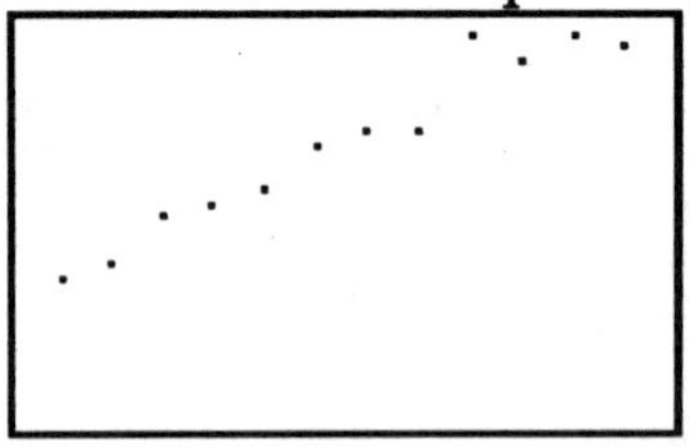

Fit a line through the average of x which is 6.5 and 1/y which is 0.008974. The slope of this line is M = 0.000663, sy 1/y = 0.000663(x – 6.5) + 0.008974. Therefore y = 1/(0.000663x + 0.004643). Graph of function and data. Range: [0, 12] x [75, 210].

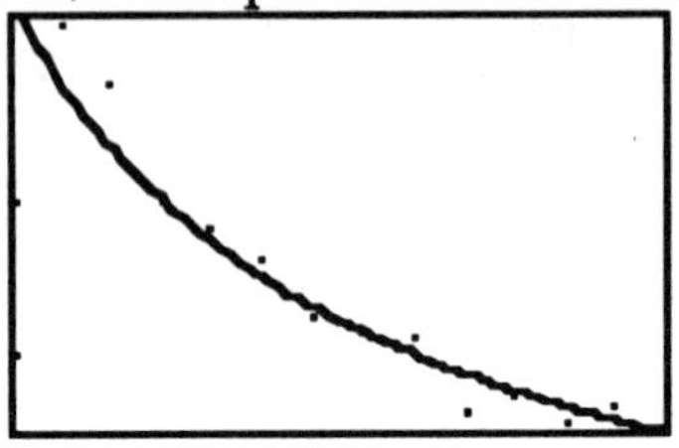

(b) At x = 13; y = 75, at x = 14, y = 72, and at x = 15, y = 69.
(c) Using the function in part (a) we find y < 45 is x > 27. So the life of the well will be about 2 years, 3 months.

Section 7.3

1. $f^{-1}(x) = (2 - x)/5$.
   One-to-one for all x in domain.
   Range: Default

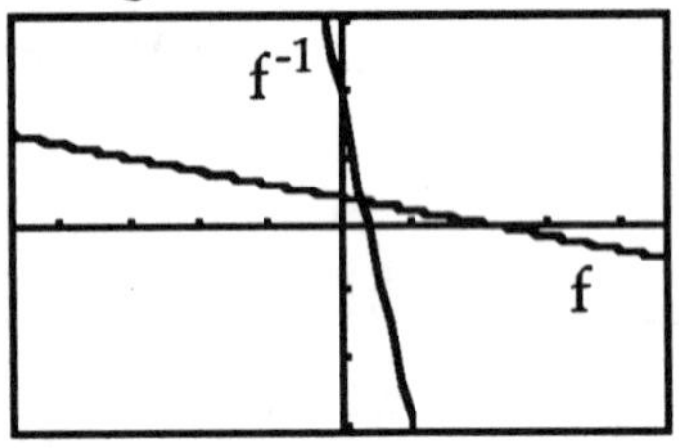

2. f(x) is one-to-one for all x..
   $f^{-1}(x) = (x^3 + 5)/2$.
   Range: [-9.4, 9.4] x [-6.2, 6.2]

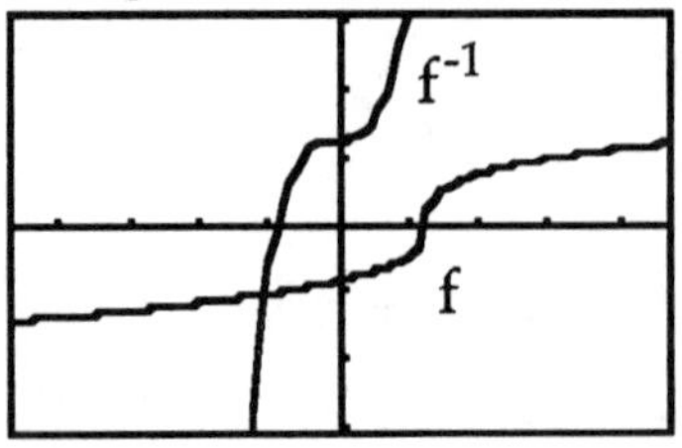

3. $f^{-1}(x) = 2x/(4 - x)$

One-to-one for all x in domain.
Range: [-23.5, 23.5] x [-15.5, 15.5].

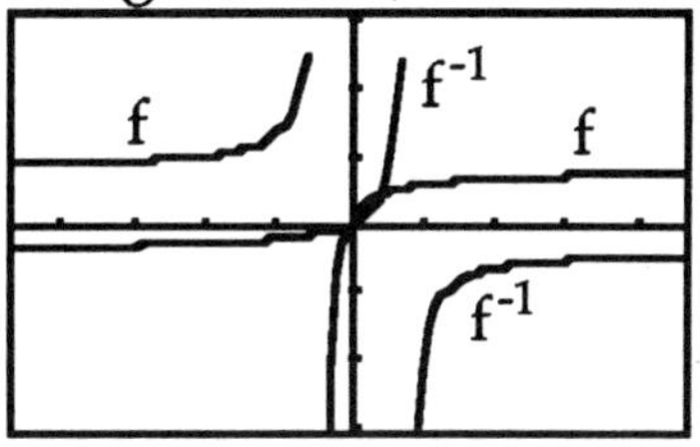

4. f(x) is one-to-one if $x \geq 2$.

$f^{-1}(x) = 1 + \sqrt{4 - x}$.
Range: [-9.4, 9.4] x [-6.2, 6.2]

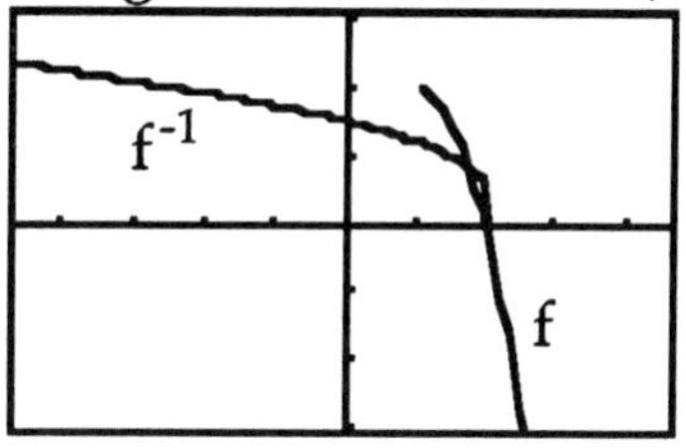

5. Range: Default

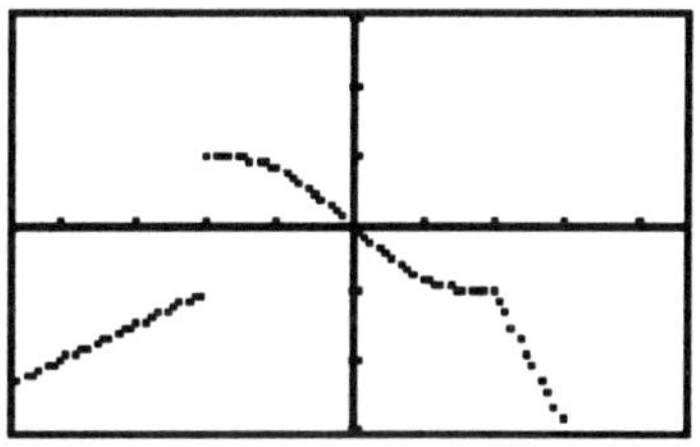

CASIO
Use Point Plotter with

```
X<-1⇒Goto 1
X<2⇒Goto 2
-2X+3→Y
Goto 3
Lbl 1:X÷2→Y
Goto 3
Lbl 2:-4X÷(X²+4)→Y
Lbl 3
```

TI-81/82
Enter this in **Y1**.

```
Y1=(X/2)(xX<-2)+
(-4X/(X²+4))(-2≤X)
(X<2)+(-2X+3)(X>2)
```

6. Default range setting.

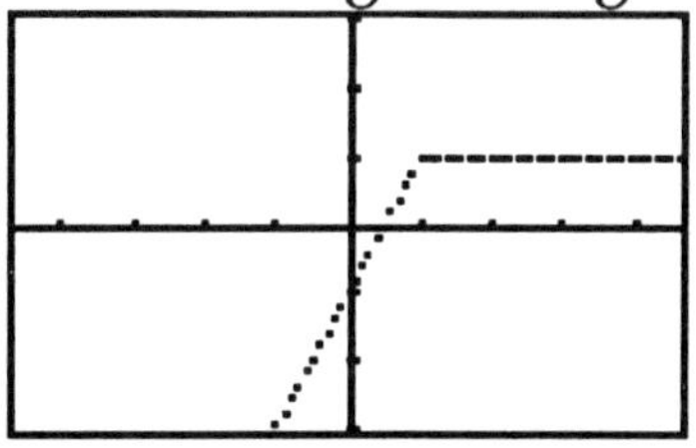

7. Range: Default

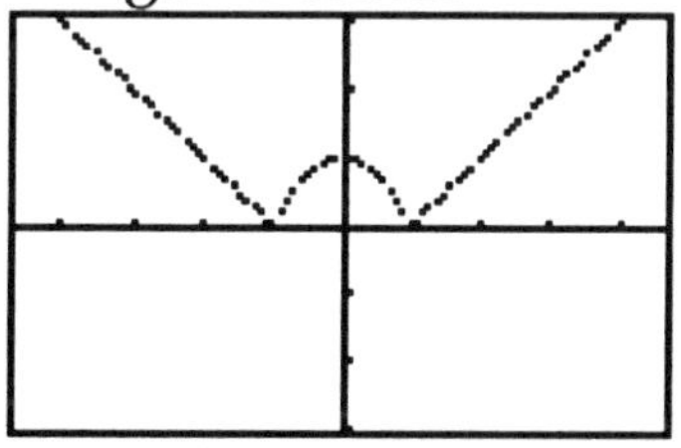

8. Range: [-2, 2] x [-.5, 1]

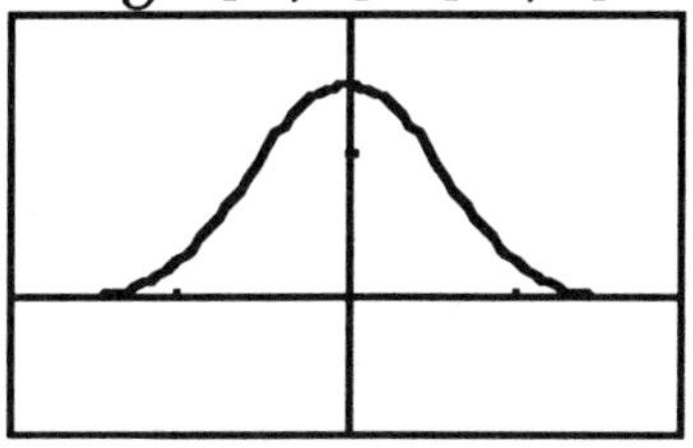

Section 7.4

1. $\sqrt{130} = 11.40175425.$
2. $\sqrt{47} = 6.403124237.$
3. $\sqrt{4 - 2\sqrt{2}} = 1.0823922.$
4. $\sqrt{2} = 1.414213562.$

5. Center: (-1, 2), Radius: = 2
Range: [-5.7, 3.7] x [-1.1, 5.1]

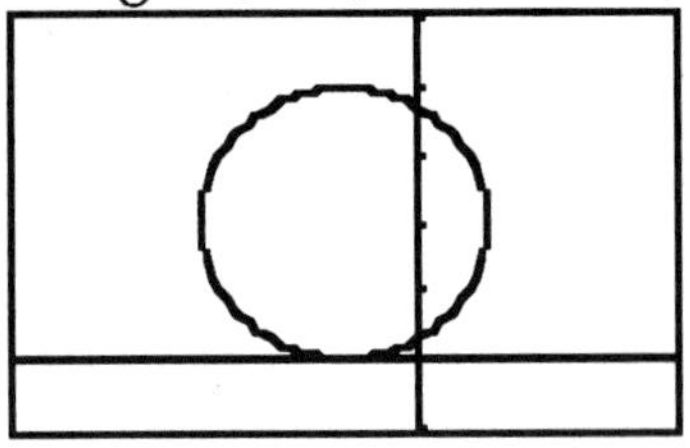

6. Center: (3/8, -1), Radius = $\sqrt{165}/8$
Range: [-5.7, 3.7] x [-1.1, 5.1]

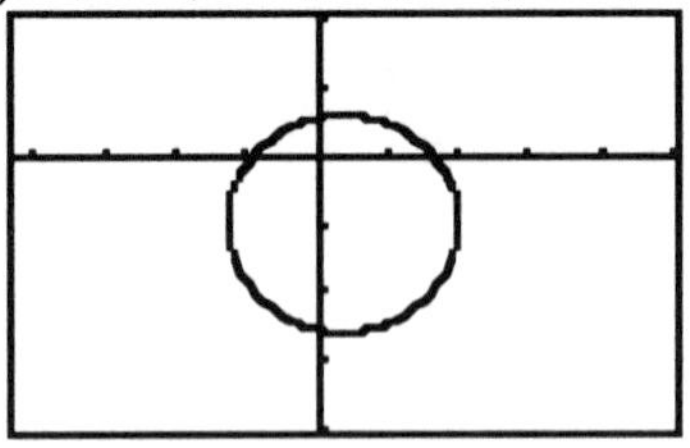

7. Center: (3, -2), Radius = $2\sqrt{2}$.
Range: [-1.7, 7.7] x [-5.1, 1.1]

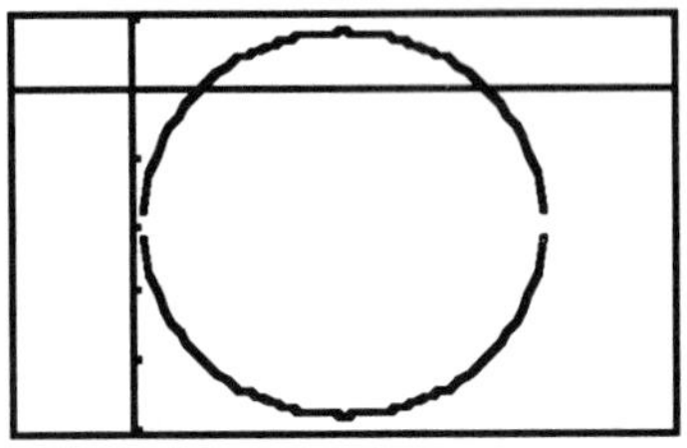

8.

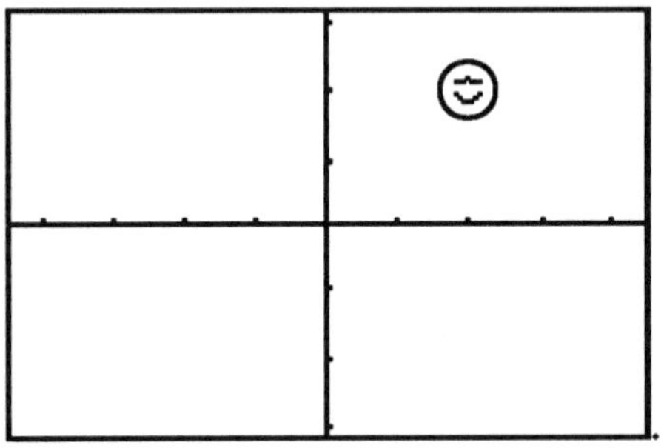

Section 7.5

For the range settings, we give the screen center, (x, y), and the scale factor, s, used in the **RANGE** setting programs.

1. $(x-1)^2/4 + y^2/9 = 1$; $a = 2, b = 3, c = \sqrt{5}, e = \sqrt{5}/2.$
Center: (1, 0); Foci: $(1, \pm\sqrt{5})$
Range: Centered at (1, 0); s = 0.1.

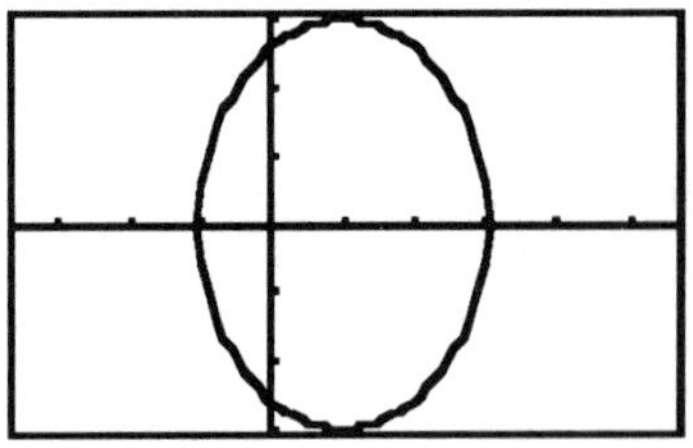

2. $(x+2)^2/5 + (y-2)^2/3 = 1$;
$a = \sqrt{5}, b = \sqrt{3}, c = \sqrt{2}, e = \sqrt{2/5}.$
Center: (-2, 2); Foci: $(-2 \pm \sqrt{2}, 2)$
Range: Centered on (-2, 2); s = 0.1.

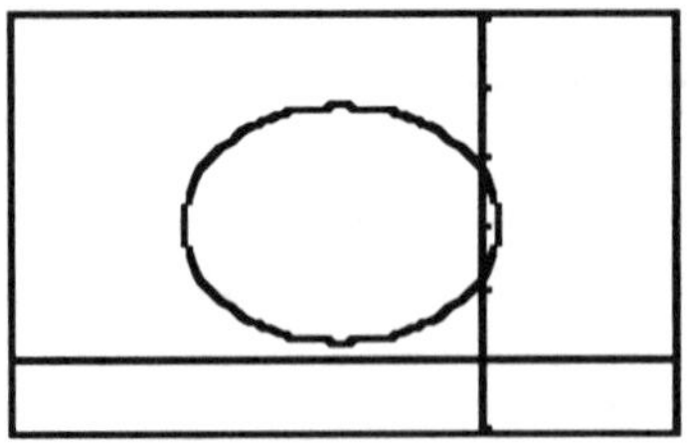

3. $(x + 1/2)^2 + (y + 3)^2/(1/4) = 1$;
$a = 1, b = 1/2, c = 1/\sqrt{2}, e = 1/\sqrt{2}$.
Center: (-1/2, -3); Foci: $(-1/2 \pm 1/\sqrt{2}, -3)$
Range: Centered at (-1/2, -3); s = 0.05.

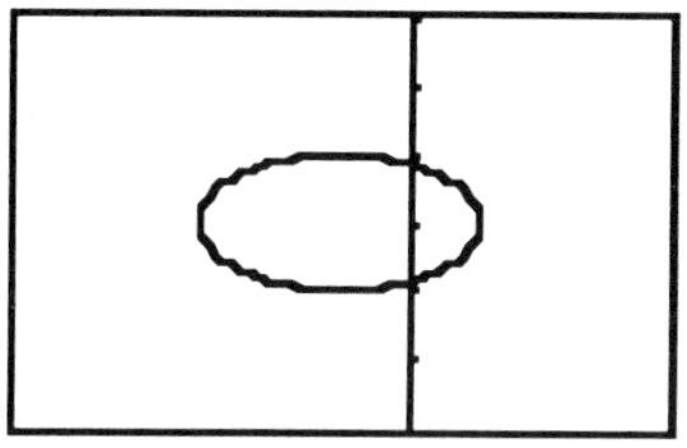

4 $(x + 4)^2/8 + (y - 2)^2/3 = 1$,
$a = 2\sqrt{2}, b = \sqrt{3}, c = \sqrt{5}, e = \frac{1}{2}\sqrt{5/2}$.
Center: (-4, 2); Foci: $(-4 \pm \sqrt{5}, 2)$
Range: Centered on (-4, 2); s = 0.1.

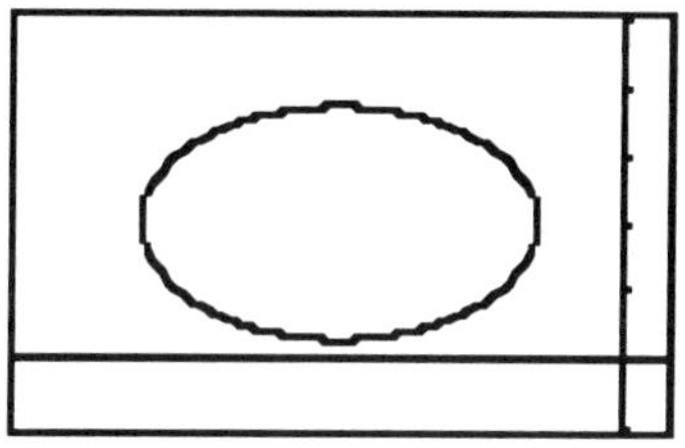

9. Orbits of first 4 planets and Halley's Comet.

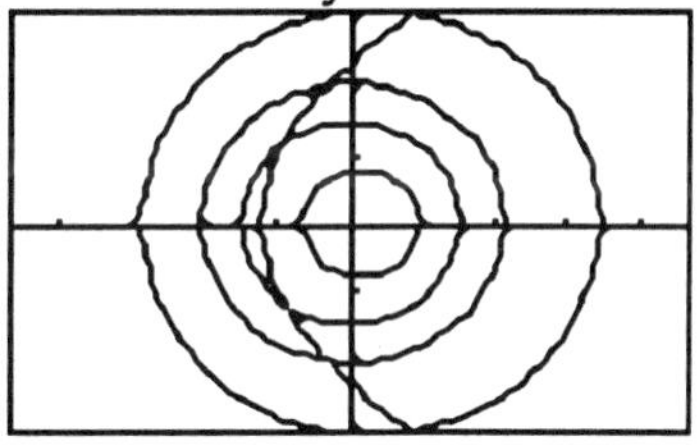

Range: Centered at (0, 0); s = 4.5.

Orbits of last 5 planets and Halley's Comet.

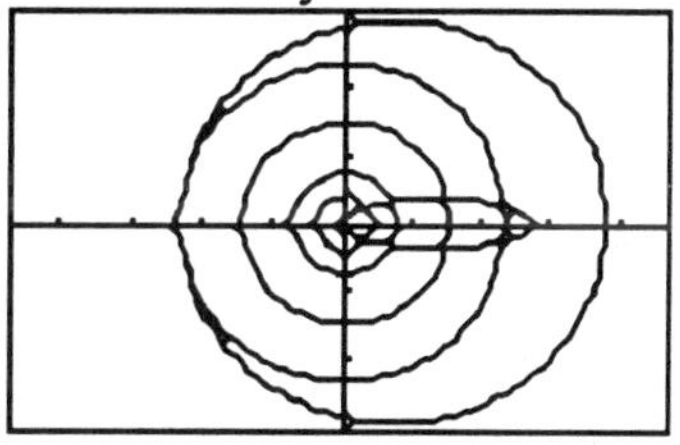

Range: Centered at (0, 0); s = 120.

Halley's Comet's orbit goes between Venus and Mercury.
Pluto nears the orbit of Neptune. In fact, Pluto's path is inside that of Neptune for part of its journey around the sun. The plane of its orbit is also different from those of the other planets.

6. $(x - 1)^2/(1/9) - y^2/(1/4) = 1, a = 1/3, b = 1/2, c = \sqrt{13}/6$.
$e = \sqrt{13}/2$, Center: (1, 0), Foci: $(1 \pm \sqrt{13}/6, 0)$
Range: Centered at (1, 0); s = 0.05.

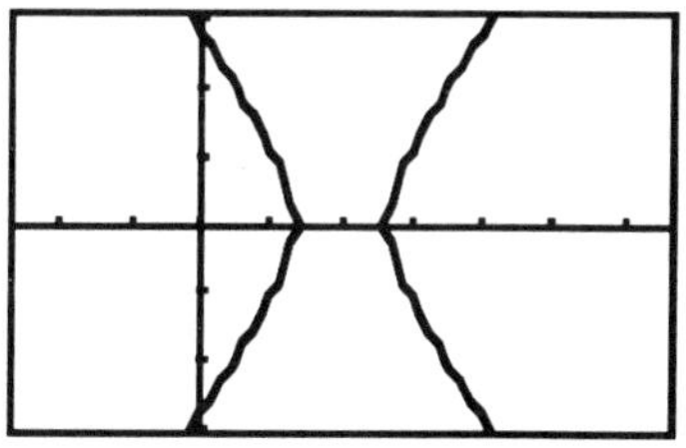

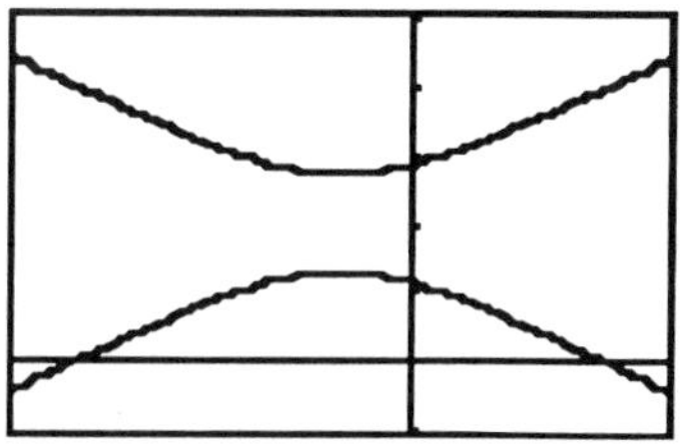

7. $(y - 2)^2/(1/4) - (x + 1)^2 = 1, a = 1, b = 1/2, c = \sqrt{5}/2$.
$e = \sqrt{5}$; Center: (-1, 2); Foci: $(-1, 2 \pm \sqrt{5}/2)$
Range: Centered at (-1, 2); s = 0.1.

8. $(x-1)^2/3-(y-1)^2/4=1$, $a=\sqrt{3}$, $b=2$, $c=\sqrt{7}$, $e=\sqrt{7/3}$.
Center: (1, 1); Foci: $(1 \pm \sqrt{7}, 1)$
Vertices: $(1 \pm \sqrt{3}, 1)$
Range: Centered at (1, 1); s = 0.1.

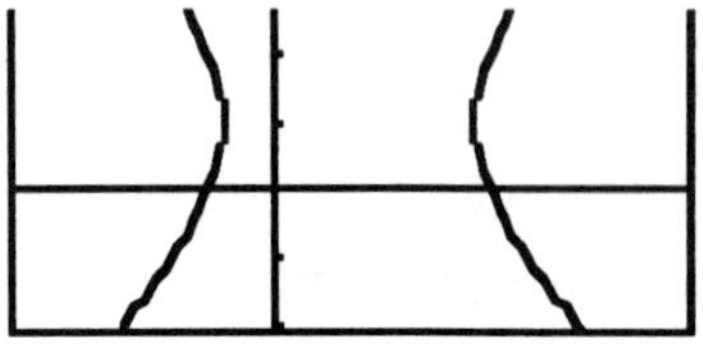

9. $x^2/8-(y+3)^2/16=1$, $a=2\sqrt{2}$, $b=4$, $c=2\sqrt{6}$, $e=\sqrt{3}$.
Center: (0, -3); Foci: $(\pm 2\sqrt{6}, -3)$
Vertices: $(\pm 2\sqrt{2}, -3)$
Range: Centered at (0, -3); s = 0.2.

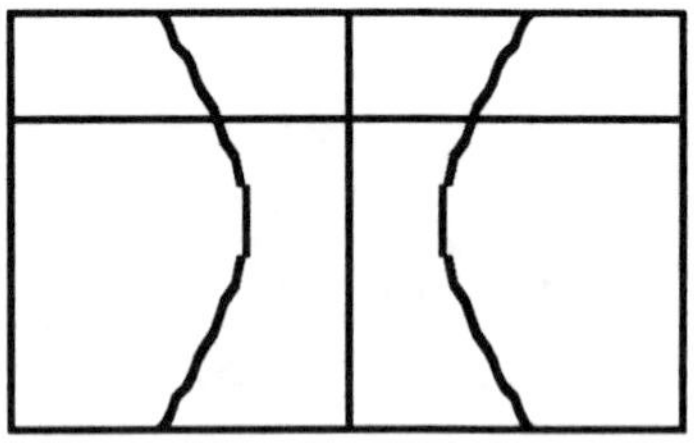

8. Assuming k > 0,then the equation is given by
$$\frac{(y-k)^2}{b^2}-\frac{x^2}{9}=1, \text{ where } b=\frac{3}{\sqrt{e^2-1}} \text{ and } k=\frac{3e}{\sqrt{e^2-1}}.$$
e = 1.5, 2, 2.5, 3, 5.

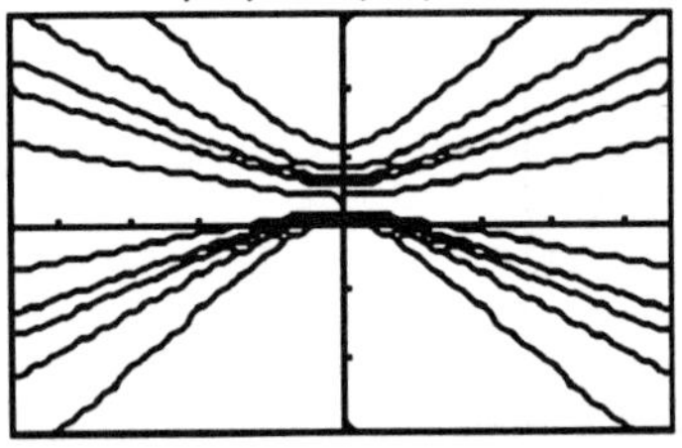

Range: Centered at (0, 0); s = 0.5.

Section 7.6

5. $x_1=4-\sqrt{12}$, $y_1=4+\sqrt{12}$; $x_2=4+\sqrt{12}$, $y_2=4-\sqrt{12}$.
6. $x=\pm\sqrt{3}$, $y=\pm 1$.
7. $x_1=-1-2\sqrt{21}/3$, $y_1=-2-\sqrt{21}/3$; $x_2=-1+2\sqrt{21}/3$, $y_2=-2+\sqrt{21}/3$.
8. $x=\pm\sqrt{2}/3$, $y=1/9$.

Answers to Investigating Inverses

To graph the inverse let $X_{2T} = T^2$, the value of Y_{1T} and let $Y_{2T} = T$, the value of X_{1T} (Figure 1). To graph y = x let $X_{3T} = Y_{3T} = T$.

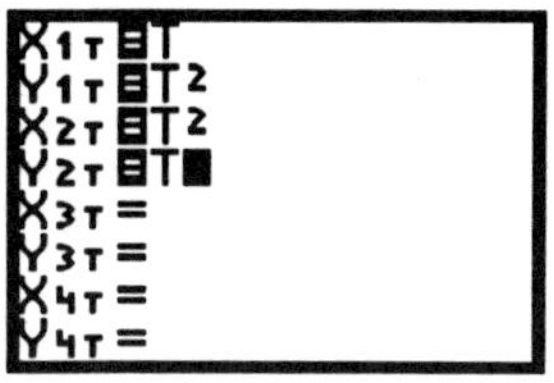

Figure 1

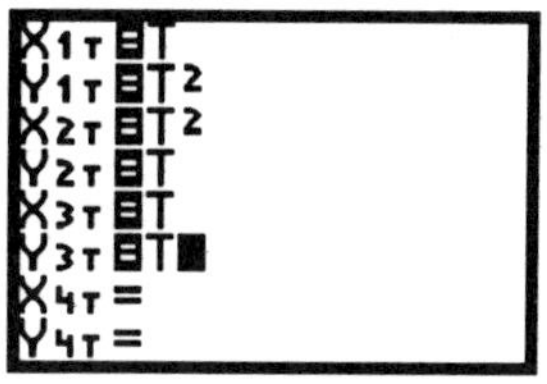

Figure 2

Original Equation	Is the original equation a function?	Is the inverse of the original equation a function?
$y = x^2 + 2$	yes	no
$y = x^4 - 2x^3$	yes	no
$y = x^2 + 2x - 3$	yes	no
$y = x^3 - 2$	yes	yes
$y = 4x - 1$	yes	yes
$y = x^5 - 2x^3 - 2$	yes	yes
$y = 2^x$	yes	yes
$y = \left(\frac{1}{2}\right)^x$	yes	yes
$x = y^2 - 3$	no	yes
$x = y^3 - 3y$	yes	yes

Table 1

For polynomial functions, if the highest power of the variable is even the inverse is not a function, but if the highest power of the variable is odd the inverse is a function. If the function is exponential then its inverse is also a function. A relation that is not a function can have an inverse that is a function.

Section 8.1

1. [-4, 2] x [0, 50]

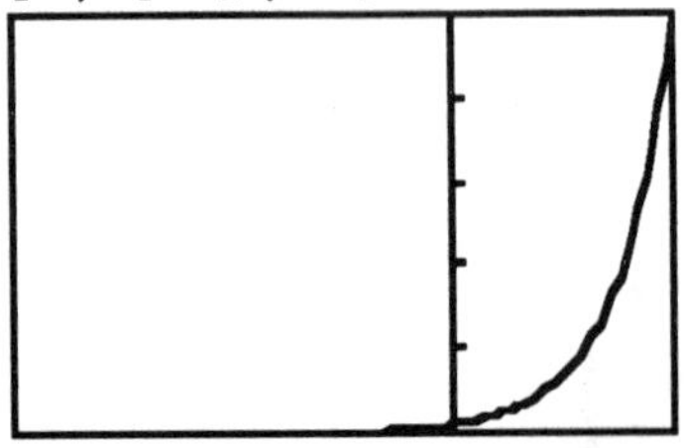

y-intercept: (0, 1)

2. [-5,5] x [0, 5]

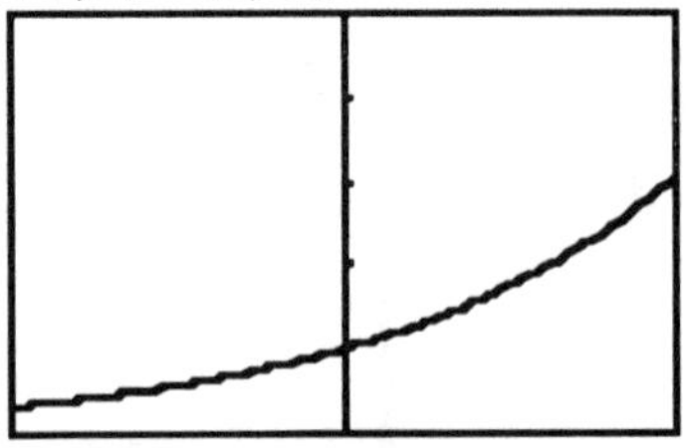

y-intercept: (0, 1)

3. [-2, 2] x [0, 10]

y-intercept: (0, 1)

4. [-6, 6] x [0, 10]

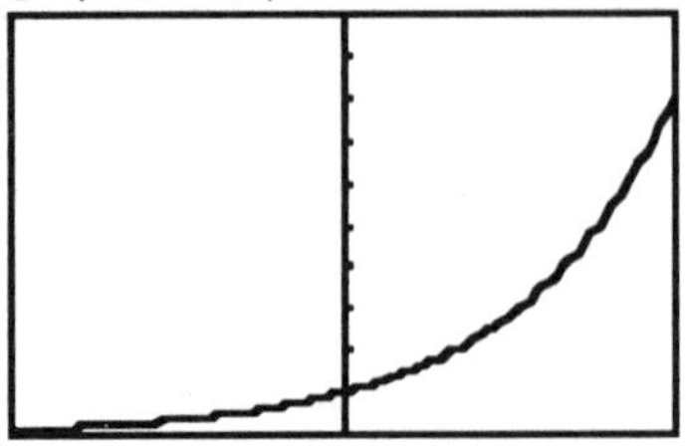

y-intercept: (0, 1)

5. [-5, 5] x [-10, 10]

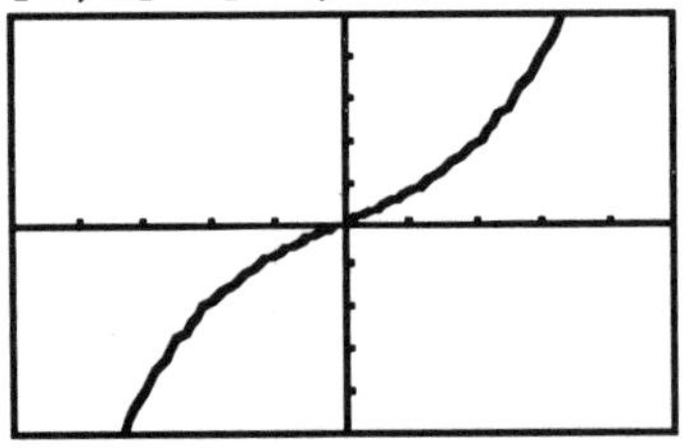

y-intercept: (0, 0)

6. [-2.5, 2.5] x [0, 1.5]

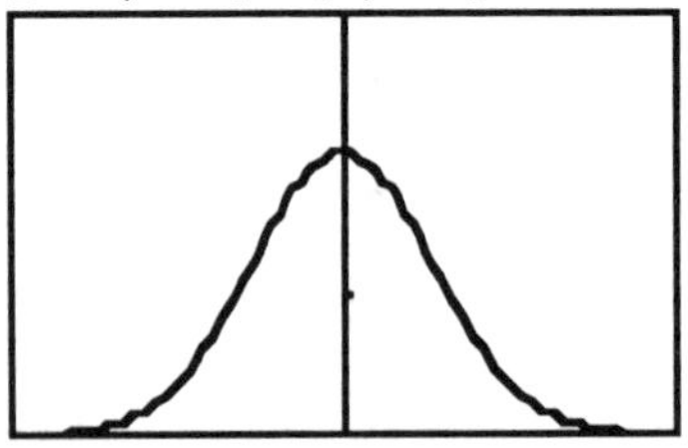

y-intercept: (0, 1)

7. $h(x) = 2(3^x)$; [-2.5, 2.5] x [0, 10]

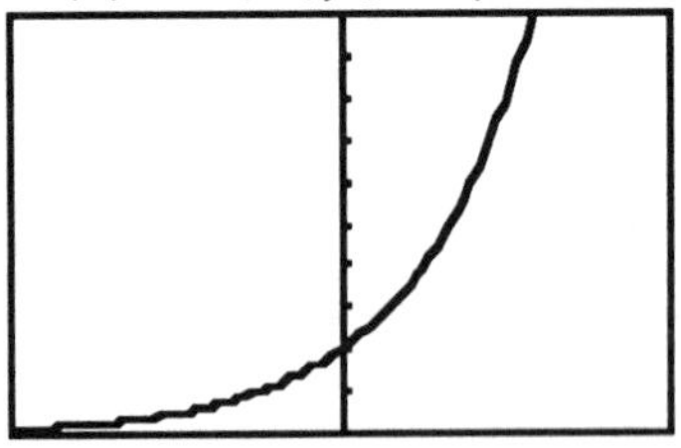

y-intercept: (0, 2)

8. $k(x) = -3(3^{x-1})$, [-3, 3] x [-10, 0]

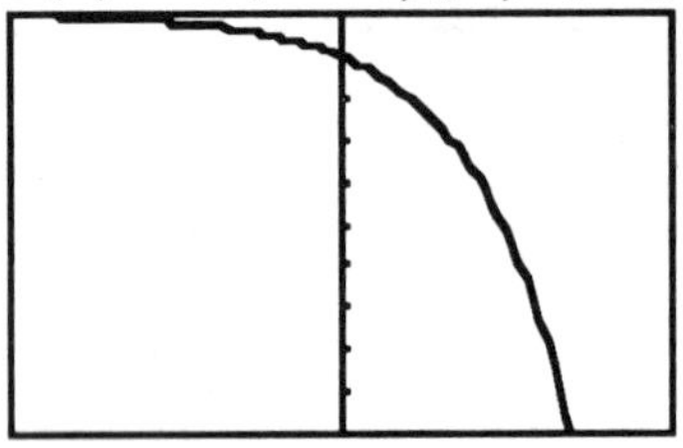

y-intercept: (0, -1)

9. $r(x) = 3^{(2x+1)} - 4$, $[-3, 2] \times [-10, 10]$

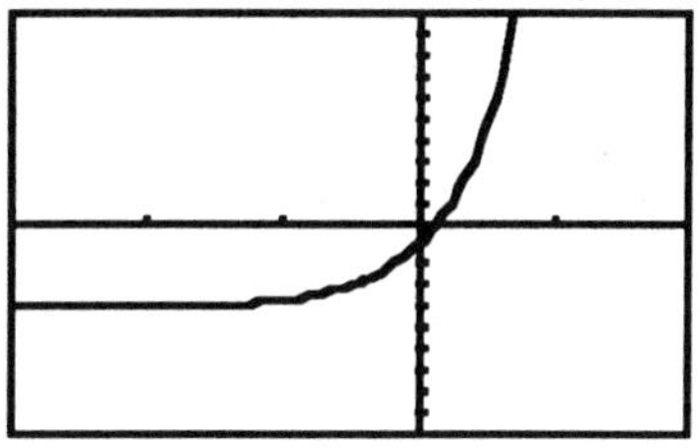

y-intercept: (0, -1)

10. $h(x) = (2/3)^x/2$; $[-5, 5] \times [0, 5]$

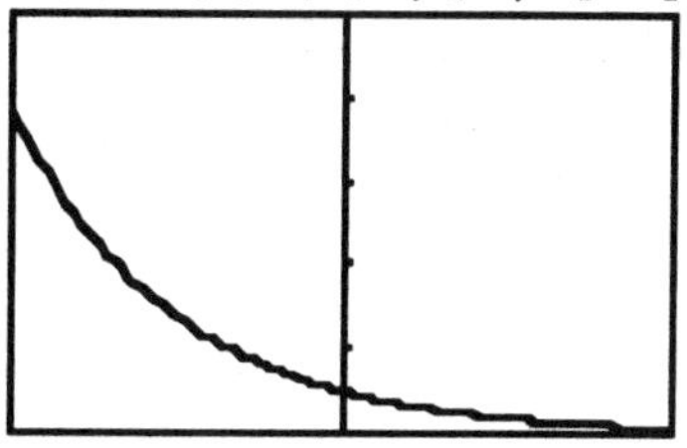

y-intercept: (0, 1/2)

11. $k(x) = -(2/3)^{(x+2)}$, $[-5, 5] \times [-5, 0]$

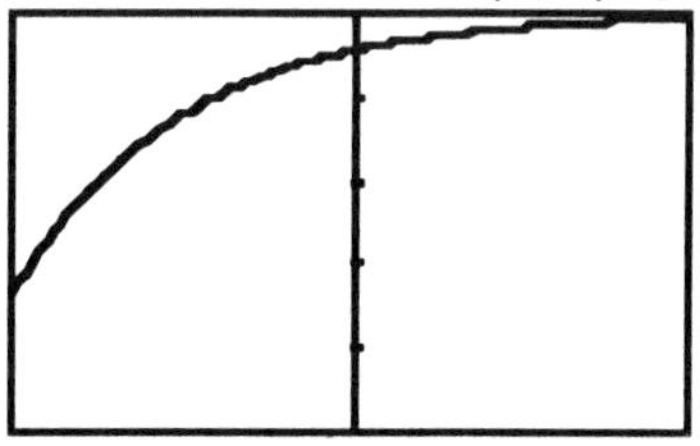

y-intercept: (0, -4/9)

12. $r(x) = (2/3)^{(4-x)} + 2$, $[-10, 10] \times [0, 10]$

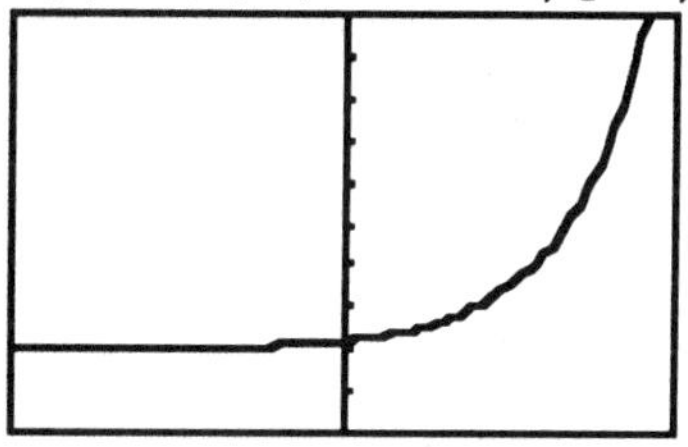

y-intercept: (0, 178/81)

13. $f(x) = (2.5)^{(2x+1)} - 1 = ((2.5)^2)^x(2.5) - 1 = 2.5(6.25)^x - 1 = h(x)$.

14. (a) $q(t) = 4.25e^{-kt}$. Since $q_0 = 4.25$ and for $t = 125$, $\frac{1}{2} q_0 = q_0e^{-125k}$, solve $2.125 = 4.25e^{-125k}$, to get $k = 0.0055451774$. Thus $q(t) = 4.25e^{-0.0055451774t}$.

(b) $[0, 625] \times [0, 5]$

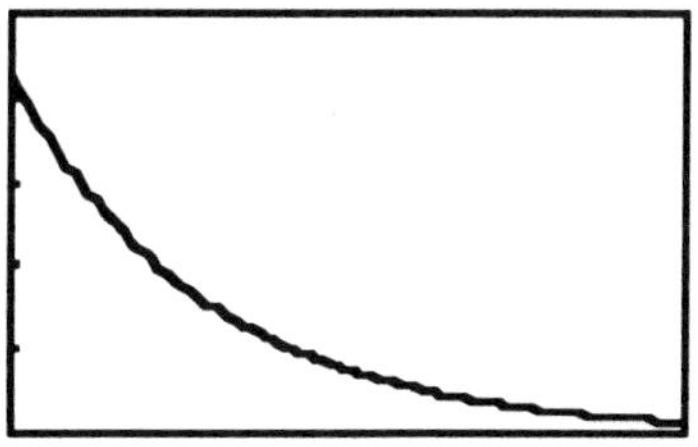

(c) Find where $y = 3.75$ and $q(t)$ intersect. $t = 22.57$ days.

15. 37072.69 years ago.

16. Doubling time is $2000 = 1000(1 + r/m)^{mt}$, or $(1 + r/m)^{mt} = 2$, i.e., $(1 + 0.085/m)^{mt} = 2$.
(a) $m = 4$, $t = 8.241012$.
(b) $m = 12$, $t = 8.183520$.
(c) $m = 365$, $t = 8.155622$.
(d) $m = 8760$, $t = 8.154712$.
(e) $m = 525600$, $t = 8.154676$.

17. Solve $e^{0.085t} = 2$, $t = 8.15467376$.
Doubles in 8 years, 56 days.
If compounded daily; doubles in 8 years, 57 days.

18. (a) f(0) = 75.99500067, f(1) = 88.84808197, f(2) = 103.4089056, f(3) = 119.7473004
f(4) = 137.8846169, f(5) = 157.7806664, f(6) = 179.323003, f(7) = 202.3207456
f(8) = 226.505004.

(b) 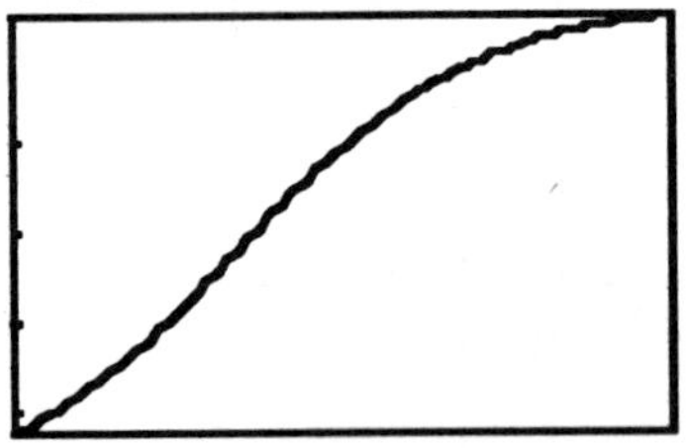

Steepest part is between x = 8 and x = 12,
The years 1980 - 2020.

(c) f(10) = 227.026, f(11) = 302.550 (d) 558.278 million.

(e) At carrying capacity, there will be about 158 persons per square mile.

(f) Solve f(t) = 406.424, giving t = 15.454, this is in the year 2054.
406.424 is about 73% of the carrying capacity.

19 (a) $h(n) = (0.004)2^n$

(b) Approximately 42 times.

Section 8.2

1. [-1, 10] x [-6, 6]

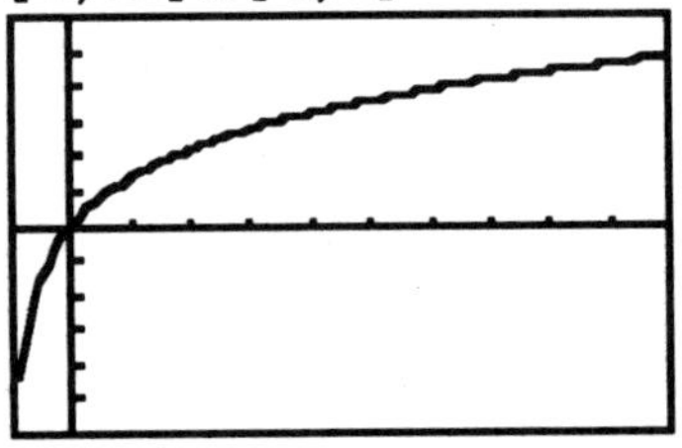

x-, y-intercepts: (0, 0)

2. [-1, 10] x [-6, 6]

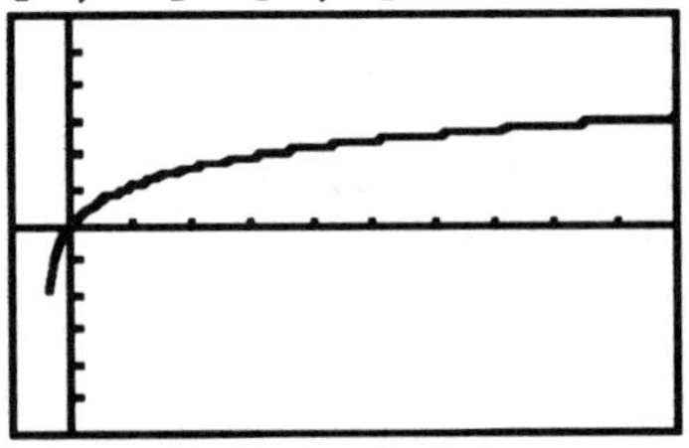

x-,y-intercepts: (0, 0)

3 [-10, 1] x [-3, 3]

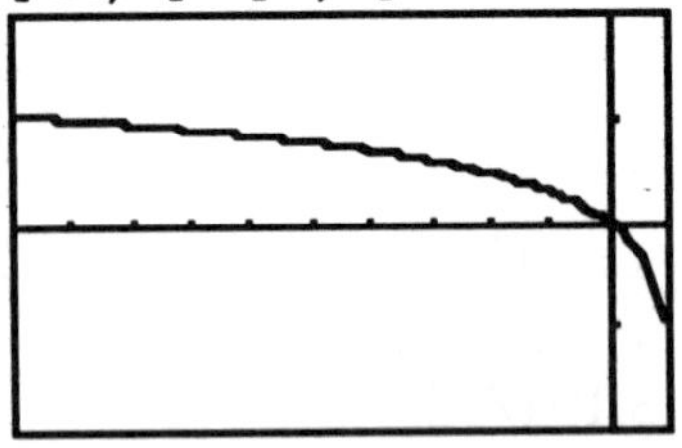

x-, y-intercepts: (0, 0)

4. [-10, 10] x [0, 3]

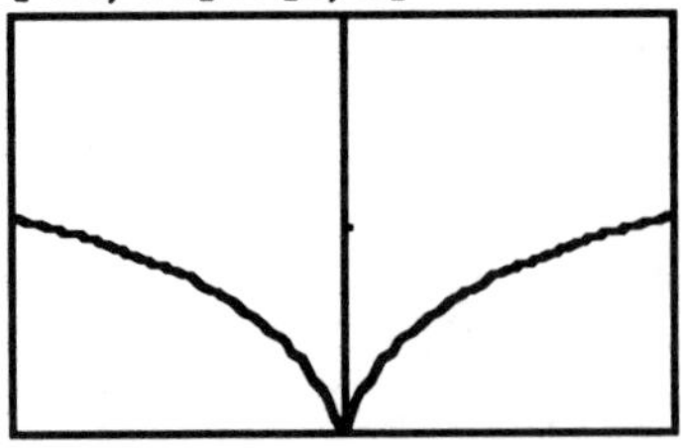

x-, y-intercepts: (0, 0)

5. [-10, 10] x [-4, 4]

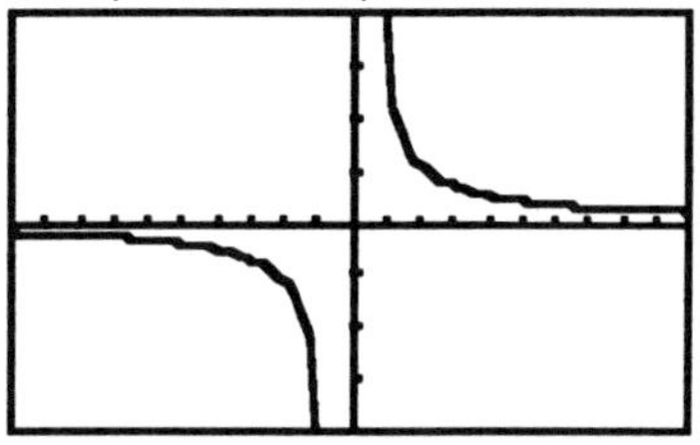

no intercepts

6. [-10, 10] x [-1, 2]

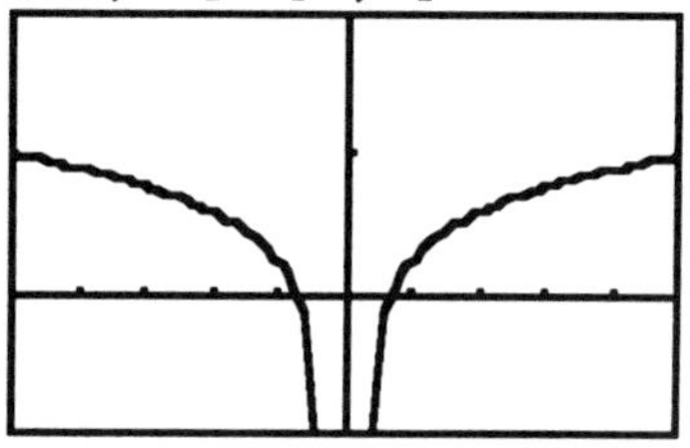

no intercepts

7. [-3, 3] x [-2, 2]

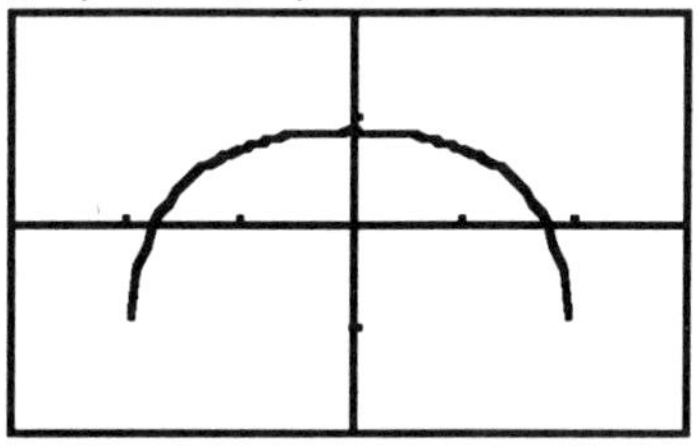

y-intercept: (0, $\log_5 4$)
x-intercepts: ($\pm\sqrt{3}$, 0)

8. [-1, 10] x [-8, 0]

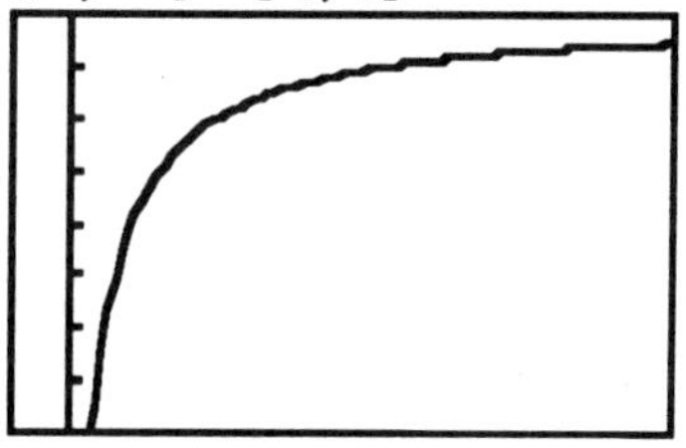

no intercepts

9. (a) The exponent m must be less than 1. If m > 1, the graph would turn the other way. Range: [$8\text{x}10^{20}$, $2.5\text{x}10^{25}$] x [4000, 140000].

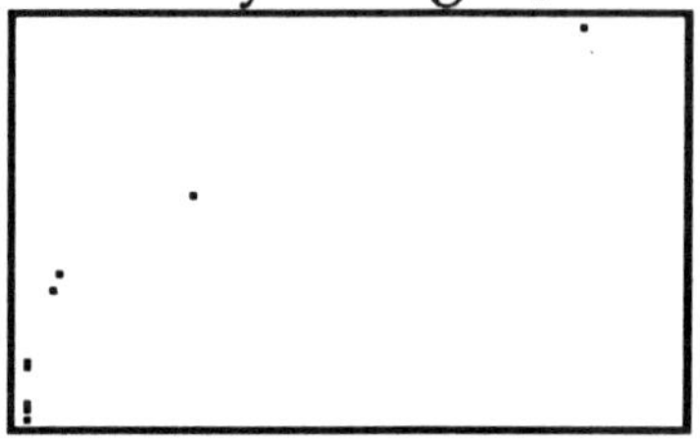

(b) m = (ln(134411.01) – ln(5291.42))/(ln($2.097\text{x}10^{25}$) – ln($8.112\text{x}10^{20}$)) = 0.3183845906.
lnV = 0.3183845906 lnM + b, so b = -6.754798938.
Range: [47, 60] x [8, 12]. Graph of (lnM, lnV) and y = 0.3183845906x – 6.754798938.

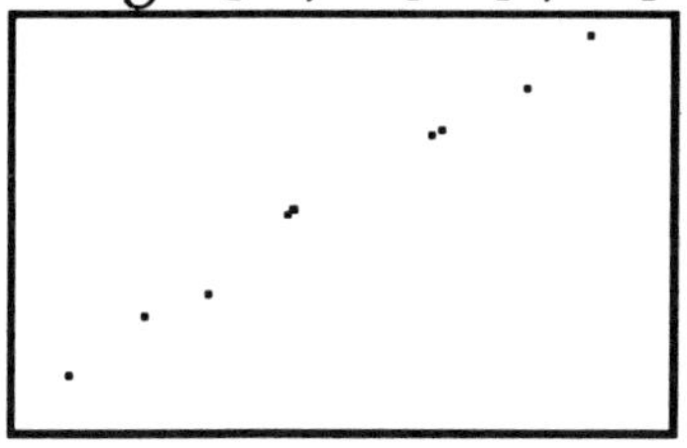

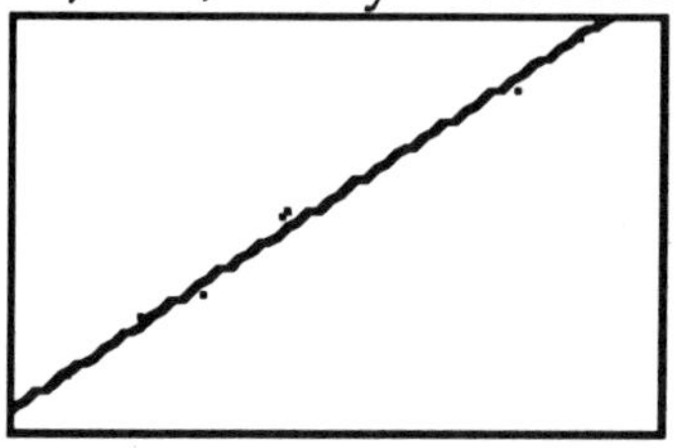

(c) $V = cM^m$ so $\ln V = \ln c + m\ln M$. $\ln c = -6.754798938$, $c = 0.0011652741$.
$V = 0.0011652741M^{0.3183845906x}$. Range: $[8\times10^{20}, 2.5\times10^{25}] \times [4000, 140000]$.

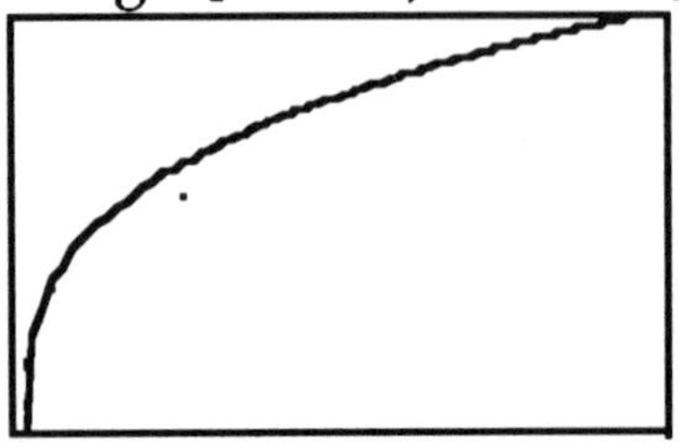

(d) Predicted value is $6166.7 = 0.0011652741M^{0.3183845906(1.312\times10^{21})}$.
(e) (R, V): [1000, 42000] x [4000, 140000]. (lnR, lnV): [6, 11] x [8, 12]

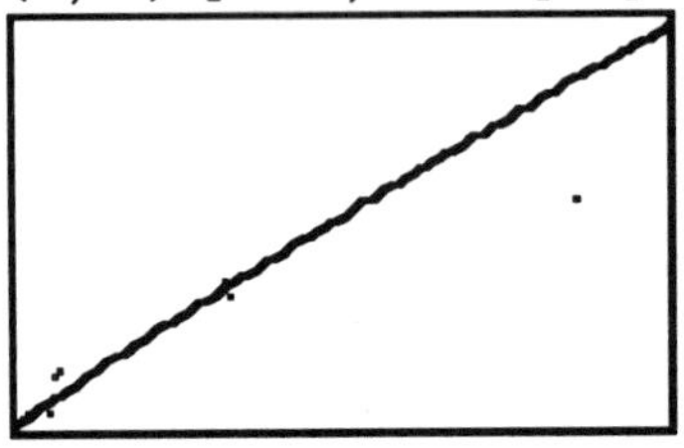

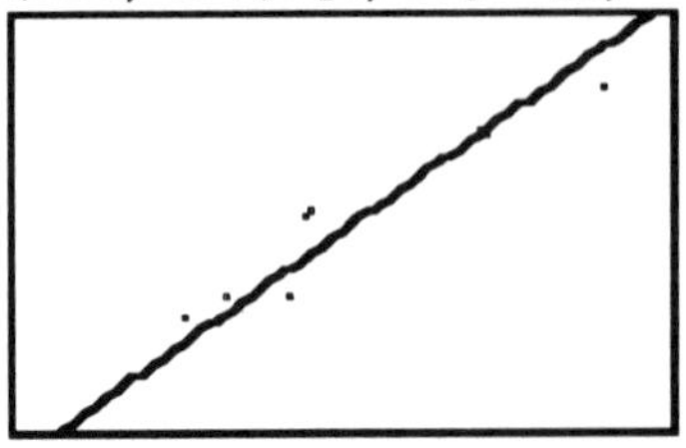

Slope: $m = 0.8864143702$, intercept: $b = 2.382571081$; $V = cR^m$, $c = 10.83271889$.
For Titan, the model predicts 6988.35. It appears that mass is a better predictor of escape velocity.

10. (a) Range [0, 10] x [4, 6]

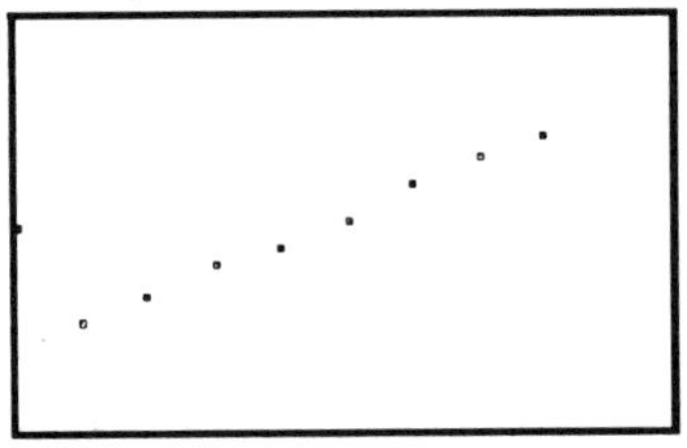

If $\ln P = mx + b$ then exponentiating both sides gives $P = e^{mx+b} = e^b e^{mx} = ce^{mx}$, where $c = e^b$.

(b) Slope: $m = 0.1365124339$, Intercept: b: 4.330667549.
So $\ln P = 0.1365124339x + 4.330667549$ and $P = 75.995e^{0.1365124339x}$.
Range [0, 10] x [4, 6] Range: [0, 10] x [75, 228].

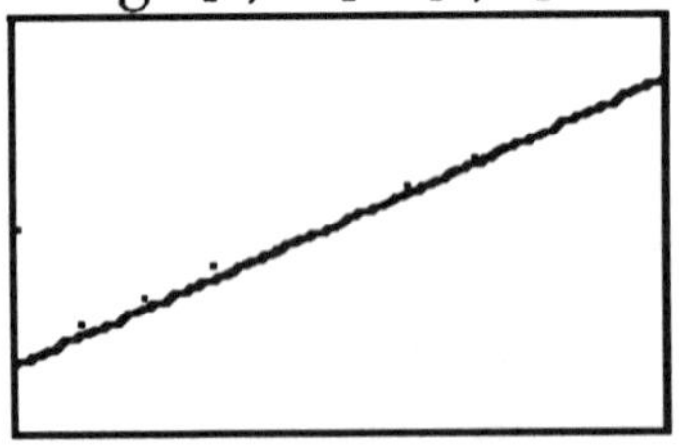

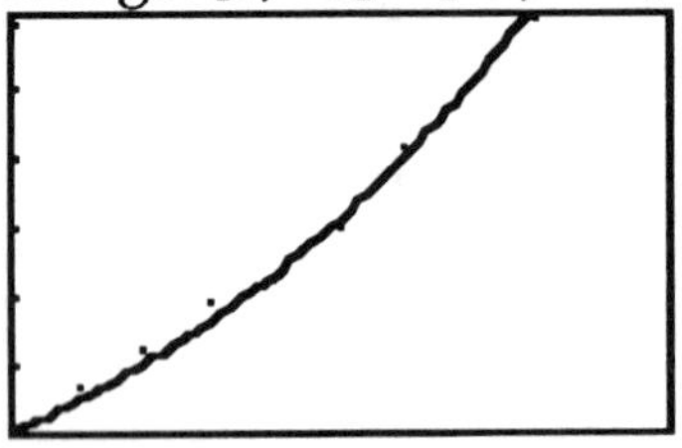

$\ln P = 0.1365124339x + 4.330667549$ $P = 75.995e^{0.1365124339x}$

(c) x = 13.078, or the year 2030. Bounded growth model, x = 18.046, the year 2080.

Answers to Investigating Exponential Growth

The information below should be included in student reports as well as other pertinent observations.

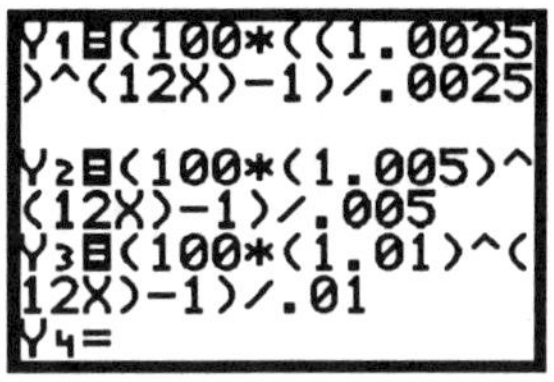

Equation Screen

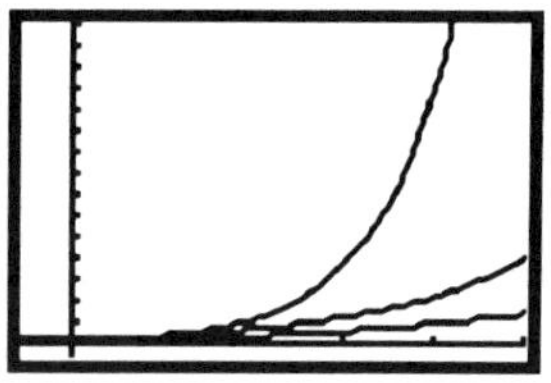

Equations

Time (years)	P (monthly payments)	Total Invested in Annuity	A_n		
			given	annual	interest
			3 %	6 %	12 %
20	\$100	\$24,000	\$32,830.20	\$46,204.09	\$98,925.54
30	\$100	\$36,000	\$58,273.69	\$100,451.50	\$349,496.41
40	\$100	\$48,000	\$92,605.95	\$199,149.07	\$1,176,477.25

Table 1

The information below should be included in student reports. Students should observe that with a constant interest rate there is a corresponding factor increase between amount of money saved and amount of annuity at a given time. For example, if you save twice as much money per month you will have about twice as much money in an annuity at the end of twenty years. However, with a constant payment rate there is not a corresponding factor increase between interest rates and amount of annuity . That is, from 6% to 12% is a factor increase of two but the amounts in annuities at the end of 40 years is a factor increase of about 3.4.

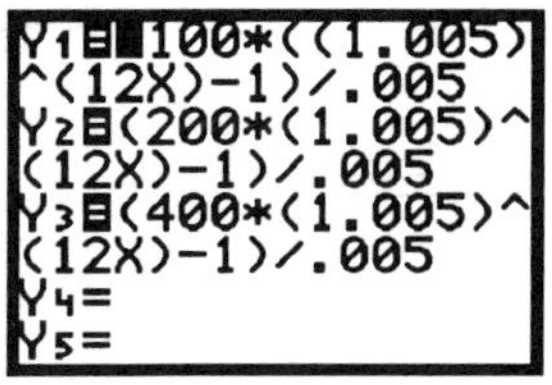

Equation Screen

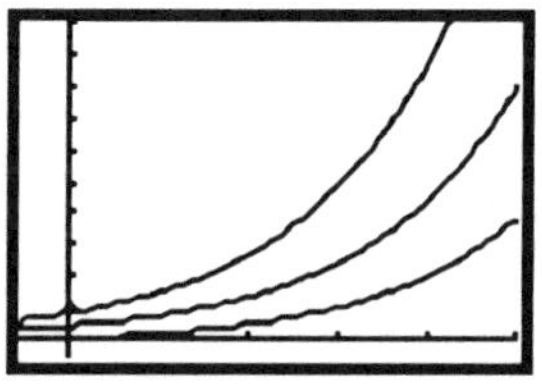

Equations

Time (years)	Interest (annual)	P monthly payments		
		$100	$200	$300
20	6 %	$46,204.09	$92,408.18	$184,816.36
30	6 %	$100,451.50	$200,903.01	$401,806.02
40	6 %	$199,149.07	$398,298.15	$796,596.29

Table 2

The answer to this question depends on the interest rate. Students who "work smart" will recognize a linear equation supplies the necessary answers. For an annual rate of 6% the twin needs to save $684.79 per month. For an annual rate of 3% she needs to save $408.01 per month. And, for an annual rate of 12% she needs to save $2,354.93 per month. Students can extend this problem by changing the age at which the twin decides to start to save money.

Table 3 shows the amount of an annuity for various time periods at 6% interest. Other interest rates can be investigated. Observe that the amount of the annuity is greater with monthly payments than with annual payments.

Interest (annual)	Time					
	20 years		30 years		40 years	
	monthly ($100)	annual ($1200)	monthly ($100)	annual ($1200)	monthly ($100)	annual ($1200)
6 %	$46,204	$44,143	$100,452	$94,870	$199,149	$185,714

Table 3

Index

A
Absolute value, 43, 61
Accuracy, 57, 58, 69
Algebraic expression, 17
Angle mode, 3, 5
Annual percentage rate, 165
Annuity, 176
Arithmetic operations, 8

B
Built-in functions, 30

C
Calculator
    basic operations
    CASIO, 7 ff
    TI-81/82, 7 ff
Coordinate system, 47
Carriage return, CASIO, 18
Circle, 147
Completing the square, 116, 120, 151
Composition of functions, 80
Conic section 130
Contrast adjustment
    CASIO, 2
    TI-81/82, 4

D
Data analysis 136
Default range settings
    CASIO 30
    TI-81/82 37
Degree mode 3, 5
Display modes for numbers 3, 4
Distance formula 147
Drawing lines 64, 65, 90

E
Eccentricity 154, 158
Editing
        CASIO 7
        TI-81/82 7
Ellipse 152
Equally scaled axes 30, 37
Equation solving 55, 68, 75, 84
Exponential functions 164

F
Factor, CASIO 32, 36
Factor Theorem 67
Factoring polynomials 66-74
Focus, foci 152, 156
Function
    exponential 164
    inverse 140
    logarithmic 170
    piecewise defined
        CASIO 143, 144
        TI-81 146
    polynomial 66
    quadratic 66, 102, 115
    rational 88

G
Goto
    CASIO 50
    TI-81/82 50
Graphics commands
    CASIO 35
    TI-81/82 41

H
Half-life 167
Hyperbola 156

I
Insertion of commands 7
Interest 165, 169, 171, 176
Inequalities 59-63, 114, 122

K
Keyboard
    CASIO 1

TI-81 4

L
Lines 55,
Linear equations 55, 90
Logarithmic functions 170
Logical tests
CASIO 70
TI-81/82 72

M
Major axis 155, 160
Matrix 126
Menus 4
Memory for number storage 14
Mode selection 2, 4

N
Natural logarithm 170
Nested multiplication 76

O
ON/OFF
CASIO 2
TI-81/82 5
Order of operation 8, 45

P
Parabola 115
Point-slope form 90
Point plotting
CASIO 47
TI-81/82 48
Program
clearing, CASIO 23
erasing, TI-81/82 27
writing, CASIO 27
writing, TI-81/82 24

Q
Quadratic formula 102
Question mark, CASIO 21
Quotes
CASIO 21
TI-81 25

R
Radian mode 2, 5
Range settings
CASIO 35
TI-81 41
Range programs 52,53
Remainder Theorem 67

S
Scale factor
CASIO 36
TI-81/82 43
Screen editing 7
Slope 90
Slope-intercept form 90
Standard form for equations
circle 150
ellipse 153
hyperbola 156, 157
parabola 115
Synthetic division 76-83

T
Trace
CASIO 35
TI-81/82 42

V
Vertical asymptote 88
Vertices
parabola 116

Z
Zoom-in
CASIO 35, 36
TI-81/82 42, 43
Zoom-out
CASIO 36
TI-81/82 43